D0153291

Biotech

POLITICS AND CULTURE IN MODERN AMERICA

Series Editors: Glenda Gilmore, Michael Kazin, Thomas Sugrue

Volumes in the series narrate and analyze political and social change in the broadest dimensions from 1865 to the present, including ideas about the ways people have sought and wielded power in the public sphere and the language and institutions of politics at all levels—national, regional, and local. The series is motivated by a desire to reverse the fragmentation of modern U.S. history and to encourage synthetic perspectives on social movements and the state, on gender, race, and labor, on consumption, and on intellectual history and popular culture.

Biotech

The Countercultural Origins of an Industry

ERIC J. VETTEL

PENN

University of Pennsylvania Press

Philadelphia

Printed in the United States of America on acid-free paper

10 9 8 7 6 5 4 3 2 1

Published by
University of Pennsylvania Press
Philadelphia, Pennsylvania 19104–4112

Library of Congress Cataloging-in-Publication Data

Vettel, Eric James.
Biotech : the countercultural origins of an industry / Eric J. Vettel.
p. cm.—(Politics and culture in modern America)
Includes bibliographical references (p.) and index.
ISBN-13: 978-0-8122-3947-8
ISBN-10: 0-8122-3947-4 (alk. paper)
1. Biotechnology industries—History. I. Title. II. Series.

HD9999.B442 V48 2006
338.4′76606 22

2006041846

To Maggie, Reed, and Whit, and for them, too.

Contents

Preface

We have had to run at full speed in order to stand still.

—Robert Glaser, October 31, 1969
Dean of Stanford University Medical School
"Message to the Biosciences"

The seemingly unlimited reach of powerful biotechnologies, and the attendant growth of the multi-billion-dollar industry, have raised difficult questions about the scientific discoveries, political assumptions, and cultural patterns that gave rise to for-profit biological research. Given such extraordinary stakes, a history of the commercial biotechnology industry must go beyond the predictable attention to scientists, discovery, and corporate sales. It must pursue how something so complex as the biotechnology industry was born, and how it became both a vanguard for contemporary world capitalism and a focal point for polemic ethical debate.

This is the story of the industry behind genetic engineering, recombinant DNA, cloning, and stem-cell research. It is a story about activists and student protestors pressing for a new purpose in science, and about politicians trying to create policy that aids or alters the course of science, and also about the release of powerful entrepreneurial energies in universities and in venture capital that few realized existed. Most of all, this is a story about people—not just biological scientists, but also followers and opponents who knew nothing about the biological sciences yet cared deeply about how research was done and how its findings were used.

There are many paths through this story, but the one followed here runs through the biological sciences at the three major research universities in the San Francisco Bay Area—the University of California at Berkeley, Stanford University, and the University of California Medical Center at San Francisco (UCSF)—during the thirty years following World War II. It is not a detailed summary of all the key discoveries that led to the creation of what is commonly known today as biotechnology, or a comprehensive study of a new scientific industry; it is a work of historical interpretation. It is a story about a young, impatient, dynamic region where people took risks to shape and then lead a scientific field. It is about the collision of culture, politics, economics, and science—that is, dramatic social and cultural change, a transforming political economy, and a sudden revolution in the biological sciences.

This is a book about the making of a biotechnology industry.

The historical narrative will follow the twists and turns of the biological sciences as they careen back and forth between pure and applied discovery. The story begins in the early postwar era when small groups of biological scientists carved a spacious and autonomous experimental niche within the larger discipline of life science. These bioscientists intended to trace the science of life to its natural beginnings, a pure science whose tributaries would converge on fundamental answers to life's most basic questions. But suddenly, in the early 1960s, a series of scientific mishaps occurred—including the thalidomide scare, the Cutter Laboratory polio outbreak, Rachel Carson's warning of permanent ecological damage—which cooled public support of unrestrained science that seemed empty of purpose. By the mid-1960s, public opinion shifted as the political right began to criticize New Deal–like government support of scientific research, while an influential political left saw pure biological research as a profound betrayal of the human side of the life sciences. By the late 1960s, the idea that bioscience research should serve the needs of people had surged through the electoral system without the calming restraint of partisan attachment, as political representatives from both parties and at all levels of government—from Lyndon Johnson to Richard Nixon, from Willie Brown to Shirley Temple Black—lent rhetorical and financial support for any biological research that had practical purpose. At the same time, a deepening economic crisis forced policymakers to slash research budgets, which left venture capital as the new resonant soulmate for biologists desperate for sustainable research patronage, even if it meant shifting experimental focus from pure to applied.

Scientists have long used terms like "pure" and "applied"—and their respective synonyms—to describe two kinds of research: the former emphasizes fundamental discovery, the latter emphasizes practical application. However, as the discerning reader probably already knows, both terms are unavoidably ambiguous and merely occupy opposite and extreme points on a continuous spectrum. Most experiments are neither entirely one nor entirely the other, especially in the biological sciences where virtually any fundamental discovery can show some practical relevance to life, and any practical application may lead to new knowledge. It is not my intention to engage an epistemological debate about the relevance of these two terms, or explain the sociological function of these terms within a dynamic scientific community. A substantial body of literature on the epistemology or sociology of experimental communities has been accumulating for some time. Rather, my goal here is to provide an account that places the travails of basic bioscience research and its corollary, the ascendancy of applied bioscience research, in historical context, and examine their relationship to the rise of the biotechnology industry.[1]

There is, in fact, ample evidence that research categories like "pure" and "applied" are historically contingent. For example, in 1967, *Science* called for an open forum in which to discuss the significance and relevance of these two terms. That a leading academic journal thought it necessary to provoke debate provides an important first clue that the meaning of pure and applied research might be historically contingent. The tone of the debate was intense. Most of the articles submitted to the journal expressed a deep revulsion with the categories and agreed that the difference between basic and applied research was often minimal and perhaps meaningless—one scientist went so far as to call it a "false consciousness." The defining features of all the essays—the hyperdefensiveness, the fierce rejection of overly simplistic descriptive categories, and the surging sense that binary categories betrayed the unity of science—offer a second clue that pure and applied bioscience research might be inextricably bound to historical context.[2]

From 1946 through the early 1960s, biological scientists at Berkeley, Stanford, and UCSF made consistent decisions about recruitment, collaboration, and publishing, and exchanged ideas between disciplines, that established pure research as superior to applied research. In general, this period was a watershed for fundamental discovery, while experiments that appeared remotely concerned with matters pertaining to medicine or agriculture were considered less worthy. That all changed in the mid-1960s when the transforming political culture and political economy compelled the opposite; many investigators responded by reconstructing their professional identities differently across time and at the same time across different disciplines. Bioscientists who identified themselves as pure or fundamental researchers in the 1950s were retroactively, in light of the rising status of utilitarian concerns during the late 1960s, eager to reconstruct their careers, laboratories, and work as part of the applied bioscience story. This shift did not merely add energy to the biological sciences; it would also release powerful popular entrepreneurial and commercial energies that few realized existed.

If there should be no meaningful distinction between basic and applied research, as none other than Albert Einstein once commented, then perhaps the same ambiguity also applies to the "biological sciences." Indeed, how should historians approach a dynamic scientific field like the biological sciences, especially during the uniquely malleable moments of the 1960s when a substantial and impressive range of fundamental discoveries were made—and then, to a measurable degree, remade, so that practical applications were paramount? Against this backdrop, what stands out about the biological sciences during this period is the consequent sweep of scientific participation and contribution. The changes taking place in the biological sciences were not occurring just in molecular biology, biochemistry, or genetics, but wherever an investigator linked life to physics and mechanics, to its chemical processes, or in anatomy, bacteriology, cell biology, embryology, endocrinology, immunology, microbiology, pathology, physiology, virology, and so on and many other subdisciplines not typically associated with the term. In other words, the reconfiguration of the biological sciences—simply, any experimental work conducted in a laboratory on a topic broadly related to life science—rather than the history of a single biological discipline is the central theme of this story.[3]

I have chosen to emphasize how this story plays in the San Francisco Bay Area—especially at Berkeley, Stanford, and UCSF. Indeed, into this booming metropolis poured unprecedented amounts of federal funds, politicized youth, and elite academic scientists. Other innovative centers, such as Route 128 in Boston, the Research Corridor in Washington, D.C., San Diego, and Seattle, or overseas (at King's College, Cambridge, and London; the Pasteur Institute in Paris; and the University of Tokyo), all housed similar arrays of constituent interests. No region, however, grew as rapidly or occupied center stage in the biological sciences and then biotechnology for as long or as significantly as the San Francisco Bay Area. Moreover, a region like the Bay Area that has three prominent universities provides an ideal site of inquiry. Consider the differences of these three universities in the context of the biological sciences: Berkeley's lack of a medical school and its success as the premier public research university restricted alternative bioscience questions; Stanford had a medical school and expanding bioscience research programs, but the combination of the two promoted intense disciplinary competition; in contrast, the focus on medical care at UCSF once meant the interests and needs of physicians superceded bioscience research, until both sides found common cause in applied bioscience research. Thus, a focus on a region rather than a single institution or discipline allows for the investigation of how a variety of actors approached evolving bioscience questions amid both historical and technical change. Put another way, a study of the biological sciences in the Bay Area offers the analytical possibility of comparative history within its own borders.

The organization of this book is primarily chronological, weaving between universities and various bioscience research programs. The first half of the book focuses on the biological sciences as they appeared at Berkeley, Stanford, and UCSF during the heyday of pure research, from the end of World War II until the mid-1960s. Chapter 1 establishes, as a point of reference, a sampling of the biological sciences in the Bay Area just after World War II. Chapter 2 offers a comparative profile of the two leading bioscience patrons of the day—the private Rockefeller Foundation and, more significantly, the U.S. government and its many constituent agencies—and then takes a closer look at *how* bioscientists were able to seize preponderant control of federal science agencies and dictate patterns of research. The narrative narrows in Chapter 3 to focus momentarily on the particular activities and organization of perhaps the single most promising bioscience program of the day: Wendell Stanley's Biochemistry and Virus Laboratory at UC Berkeley. Disciplinary tension in general, and the destructive consequences of a program rigidly committed to basic research, is examined in the context of rapid scientific changes—notably, the discovery of DNA's double helix and its internal copying mechanism. Chapter 4 steps back to show how a new group of bioscientists and university administrators—led by, among others, Arthur Kornberg's laboratory at Stanford, Julius Comroe's Cardiovascular Research Institute at UCSF, and Donald Glaser's program in molecular biology at Berkeley—moved in unison and without hesitation toward an unwavering commitment toward fundamental research.

The second half of the book explores in greater detail the relationship between bioscientists, society, and the state, chiefly by looking at the waning popular and financial support for basic bioscience research. Chapter 5 explores how activists in the Bay Area reacted against the idea of basic research and challenged bioscientists to rededicate their work toward more practical concerns. Chapter 6 examines a federal policy realignment that actively promoted practical bioscience research objectives over pure. The issues in this chapter are motive and representation, in a highly political sense—government officials responded to the public's distress by implementing policies that encouraged greater commitment by investigators to practical bioscience research objectives. Chapter 7 shows how the bioscience community, fraught with its own internal divisions and disciplinary competition, struggled against and then accommodated the shifting political culture and political economy. This chapter concludes by identifying a wide sample of applied bioscience research projects, including the development of several bioengineering techniques in the laboratories of Paul Berg and Stanley Cohen at Stanford University and Bill Rutter and Herbert Boyer at the University of California, San Francisco. In Chapter 8, the narrative focuses entirely on the birth of Cetus Corporation, the world's first biotechnology company. In this chapter, a doctor, a biologist, a physicist, and a venture capitalist break off from traditional academic models to exploit untapped commercial potential in the biological sciences. This book concludes by looking at the desperate response of biological scientists to ideological pressure, weakening of federal science policy, and privatization of research, and how their collective response inadvertently fueled the biotechnology industry.

In writing this book I have drawn on a rich body of sources. Fortunately, the biological sciences are exceptionally well documented in the Stanford, UCSF, and Berkeley archives, and in places far from the San Francisco Bay Area, including the University of Chicago, the Rockefeller Archive Center, and the Smithsonian Institution, National Museum of American History. I was also very fortunate to have found an obscure collection of primary sources held at the Pacific Studies Center in Mountain View, California, and generously given access by Chiron Corporation in Emeryville, California, to the private collection of papers documenting the historic rise of Cetus. I owe a deep debt of gratitude to executives at Chiron, for they understand better than most the importance of primary archival research. I also spoke directly to many of the historical actors in this story, and I used the vast collection of oral histories produced by the Regional Oral History Office at the University of California, Berkeley. Wherever possible, I have intentionally allowed the historical actors to speak in their own voices, providing an evocative portrait of a fast-paced scientific field, an even faster-paced industry, and a people caught in a revolution they only partially understood at the time. I am aware that individual accounts of the past rarely coincide with one other. Participants have only a partial view of events as they unfold, and over time, memories fade. Moreover, many actors have a peculiar interpretation of the past—for instance, left-leaning investigators blamed the Nixon administration for slashing support for basic bioscience research, and more conservative investigators pointed accusing fingers at the misplaced agendas of student radicals. The profusion of available evidence—oral histories, interviews, periodicals, university administrative records, corporate archives, scientific notebooks and publications, and so on—meant that different "facts" could be compared and checked against other documented sources. I am grateful to have had access to these resources, for they provided me with much needed distance from the many contentious debates that surround biotechnology and the industry.

Among the important implications of this account is a corrective that broadens the matrix within which we think about what it means to do science, and thus challenges those who naively celebrate or lament the power of scientists, patrons, or activists. This story provides essential historical background for contemporary debates on bioethics, genetic engineering, gene ownership, and cloning. The arguments contained herein also apply to other technical fields and disciplines; all science crosses the threshold into society, and is unavoidably reshaped once it gets there. The biological sciences may have been unusually dynamic in the San Francisco Bay Area—it is arguably the epicenter of the biotechnology industry—but its history shows the undeniable relationship between science and society. The lessons from this story should make possible a better measure of the biotechnology industry, then and today.

Chapter 1
The Setting, 1946 . . .

Do thou but choose, oh Noble Sirs,
For 'tis as sure as Fate
Thy deeds done of this day
Shall light thee down
Time's pathways of the future,
To Fame or Infamy
Do thou but choose.

—Herbert Evans, bioscientist at the
University of California,
regarding the future of the biological sciences,
from the vantage point of 1946

The San Francisco Bay Area, with its own history of scientific successes and its emerging presence at the edge of a technological revolution, had established itself by the end of World War II as a preeminent research center. Here, some critical discoveries had occurred, particularly in wartime production industries—radar, microwave, communication, electrical engineering, and computation. Seemingly every day throughout the war someone announced a stunning development: the Varian brothers invented a radar system; Berkeley physicists split atoms and released millions of electron volts of energy; Charles Litton gave up glassware manufacturing to build vacuum tubes; Bill Hewlett and David Packard invented a series of gadgets that produced controllable and accurate electronic signals. And because of the actions of Bay Area scientists and engineers, an infrastructure took shape, with the federal government, business, and higher education each occupying one leg of an increasingly intricate and powerful relationship. Extending the network still further, as the Bay Area grew as a research center, so too did the needs of the region, only to be fulfilled with the arrival of more scientists, machinists, managers, technicians, and engineers.[1]

Yet, despite irrepressible enthusiasm for anything scientific and technical, the Bay Area was a virtual backwater in the biological sciences.

To be sure, the Bay Area had its share of private biological firms: among others, Stayner, Lederle, and Abbott Laboratories in the East Bay, and divisions of Cutter Labs, Sharpe & Dohme, and DuPont in and around San Francisco. Caught in a great rush to duplicate the wartime antibiotic successes—such as the sulfas, penicillin, and streptomycin—these local companies focused entirely on production of pharmaceutical agents, competitive pricing, and efficient distribution processes, but typically neglected research. Companies like Stayner Laboratories sent soil microbiologists to every corner of the world to sift through samples of dirt for the next "miracle mold," Cutter Labs chemists tried to modify fermentation processes to increase polio vaccine yields, Lederle Labs grew and sold biological cultures, Abbott Labs provided simple screening services to test the potency of biological agents, and larger pharmaceutical companies continued to extract hormones from pig and cow cadavers rather than search for substances safe for human use. One divisional laboratory manager, noticing the general trend toward production and away from research, complained that everywhere he looked he saw "enormous sums of money invested in chemical production" while "chemical effectiveness was rarely understood."[2]

More curious than the general ineffectiveness of the local biological industry, the biological sciences at the three major universities in the Bay Area—University of California, Berkeley; the University of California Medical Center in San Francisco; and Stanford University—languished with feeble experimental output and unexceptional student enrollments. Certainly each campus had isolated successes. During the 1920s and 1930s, the UC Medical Center worked closely with the San Joaquin Valley's canning industry to help prevent botulism outbreaks, UC Berkeley biochemists identified and isolated a number of vitamins in pure form, and Stanford biologists "did more for U.S. fisheries than any other educational institution in the United States." However, to anyone other than the most provincial, the productivity of biological scientists in the Bay Area by the end of the war can only be measured as a collective disappointment.[3]

Bioscientists in the Bay Area, then, probably interpreted the scientific successes during the war much like everyone else, with the same jumble of ambivalent feelings: joy and relief, doubt and fear, a sense that perhaps science could one day go too far. But for them, the growing importance of scientific research and development promised, if not dramatic experimental results, every reasonable expectation to push their programs in the immediate future. Experimental biologists at all three Bay Area universities, saddled with a history of remarkable scientific achievement to which they contributed very little, understood they now had before them new opportunities.

Three universities in the postwar era: Each molded by World War II, each destined to become leading academic institutions, each determined to play a leading role in the biological sciences. Of the three, only UC Berkeley was recognized internationally as a great university, but its administration wanted more. In the 1910s, just four decades after inception, Berkeley had catapulted into the Big Six of elite universities, by the 1920s had moved into a tie for second place with Chicago, Columbia, and Yale, and in the 1930s held sole possession of second place, behind only Harvard. Then, just as suddenly, Berkeley's rapid advance slowed to a standstill. Despite the growing number of Nobel laureates, its legendary contributions to the war effort, and the greatest number of top departmental rankings for any individual university in the country, Berkeley could not supplant the forebear of higher education. University administrators agonized about the ranking that lay just beyond their grasp. How could Berkeley break through the habits of tradition? What sort of strategies might shake up these rankings and at the same time best serve Berkeley's interests? In short, what must the university do to supplant Harvard as the best university in the country?[4]

Berkeley's near-ideal showing posed two overarching problems for those who cared about these sort of things: how could Berkeley seize the highest spot with so many academic programs already having achieved elite status, and conversely, which programs hurt their overall ranking and what could be done about them? It was in this context—of academic rankings, both real and imagined, in the driving ambition to satisfy sincere intellectual curiosities, in the pursuit of scientific distinction, and in the fickle hierarchy of shifting academic reputations—that Berkeley faculty and administrators identified the biological sciences as in greatest need for renewal. It was a wise choice: many critics agreed with one published report of 1947 that "while certain aspects of science . . . perform exceedingly well, . . . this great University is as yet weak in biological sciences." Depending on the poll, the biological sciences at the University of California ranked somewhere in the range of twenty-fifth to thirty-fifth in the country. The biological sciences at the University of California were arguably the weakest academic program.[5]

President Robert G. Sproul determined to strike boldly. A graduate of Berkeley, Sproul had spent just one year away from his alma mater working for the nearby city of Oakland, and then promptly returned to the university where he worked for the next forty-four years, twenty-eight as its president. Parochial he may have been, but he had many of the right qualities to lead a public research university such as Berkeley: an extraordinary dedication to managing details, the ability to take considerable pressure, and, if need be, the courage and political acumen to force through unpopular decisions. Hopelessly unwilling to trust anyone but himself with his dream to build Berkeley into the country's "leading academic institution," Sproul moved quickly to solve Berkeley's problem with the biological sciences.[6]

That fateful opportunity came in 1946 when Berkeley's most prominent and unquestioned leader in the biological sciences, the biochemist C. L. A. Schmidt, fell ill and died. While colleagues mourned his passing, President Sproul led a behind-the-scenes search for a replacement "who could shape an enlightened national and international biological research agenda." During a crosscountry trip, President Sproul "fortuitously" bumped into Wendell Stanley—a biochemist of world renown—and shared his frustrations.[7]

The president's clandestine overtures precluded any immediate interest from Stanley, for the simple reason that Stanley saw Sproul's grand dream and Berkeley's weakened state in the biological sciences as "largely incompatible notions." What was more, Stanley knew that Sproul's lavish expectations would flummox the current faculty in the biological sciences. For years, Sproul had allowed the biosciences to expand virtually unrestrained across the full terrain of the field. While physics thrived under such liberal administrative conditions, the biosciences had become, by 1946, less organized, less productive, and more isolated than the ideal of free science allowed. Nor did it necessarily work automatically. Though the exact number of faculty working in the biosciences was conjectural, Sproul was in essence asking Stanley to hold together more than one hundred full-time faculty in twenty-nine experimental programs scattered throughout the Arts and Sciences, the College of Agriculture, and the Pre-Clinical Sciences. Clearly, it was not simple administrative timidity but practical realism that had choked Berkeley's rise.[8]

Stanley, in the habit of speaking bluntly, presented Sproul with a specific solution to improve Berkeley's biological sciences. First, every bioscientist at Berkeley must focus all of their professional energies on a single, highly specific research topic (much like he had focused on virology) around which a top-flight chairman (much like himself) had full charge to trim "deadwood" staff and replace them with hand-picked specialists. Furthermore, and in Stanley's mind most crucial, the remaining faculty must sever their connections with medical or agricultural research and unite as a free-standing or autonomous department, dedicated entirely to pure research.[9]

Ever receptive to a novel plan, Sproul easily succumbed to Stanley's seductive logic. Not least among Sproul's reasons for backing the initiative were Stanley's professional aura and dogged determination, which seemed perfectly suited to the leader of a first-class biological research program. To a degree uncommon among Berkeley's biological scientists, Stanley appeared to be a steadfast scholar, even something of an academic opportunist. Ironically, Stanley's impatience and unwavering self-confidence that Sproul appreciated would eventually become his greatest burden as an academic leader at Berkeley. So would his experimental focus, established early in his career and reinforced by early successes.

Stanley, forty-two when he met Sproul on that auspicious flight in 1946, had moved swiftly through his scientific training to become, in his own unapologetic words, "just about the top experimentalist in the country." A colleague once described him as "endowed with vast skills that matched the impressions that he had of himself." Stanley's air of superiority probably came from continuous educational successes. He graduated from Earlham College in 1926 with degrees in chemistry and mathematics, raced through the doctorate program in chemistry at the University of Illinois at Champaign-Urbana, and then in 1929 won a National Research Fellowship to study chemistry in Munich—perhaps the strongest chemical research center in the world. Two years later, in 1931, Stanley accepted an appointment that changed his life: biochemistry research at the Rockefeller Institute for Medical Research at Princeton University.[10]

While most investigators at the Rockefeller Institute studied the treatment of disease, Stanley wanted to understand the process of disease, to spend more time "at the lab-bench rather than in the hospital," where he could apply his training in basic chemistry without the distraction of patient care. It was during Stanley's time alongside clinical researchers that he developed an acute and near maniacal interest in viruses. This interest made him an enthusiastic proponent of virology at a time when most experimental topics were still prescribed from traditional medical and agricultural concerns.[11]

To understand the viruses, Stanley determined to do something no one else had done before: he wanted to see them. All day and through many nights Stanley holed up in his lab and took photo after photo of crystallized TMV (tobacco-mosaic viruses) and pecked out his impressions on a portable typewriter. Soon his pictures and observations started to appear on colleagues' desks and in dozens of journal publications. It took Stanley nearly two years to convince a skeptical scientific community to accept his hypothesis that viruses were identifiable and self-reproducing proteins. *Scientific American* touted Stanley's crystallization of the virus as "unbelievable" and "wholly novel," while prominent scientists described the work as "the most important breakthrough in understanding the molecular basis . . . of biology." That was not nearly ambitious enough, said another: "No discovery made at the Rockefeller

Institute, before or since, created such astonishment throughout the scientific world." The popular press also gave Stanley's work a great deal of attention; the *New York Times* credited Stanley with unraveling the riddle of life: "in the light of Dr. Stanley's discovery, the old distinction between life and death loses some of its validity." Stanley's continued scientific accomplishments, including the development of an important influenza vaccine during an epidemic that had slowed military recruiting during World War II, added to his already considerable prestige.[12]

Sproul could believe without difficulty that Stanley's accumulating achievements, as well as his experimental focus and approach, were ideal for an elite research university such as Berkeley. For these reasons, Sproul unashamedly approached the University of California Regents and proposed an ambitious vision of an enormous, state-of-the-art 30,000-square-foot laboratory for the biological sciences, with Wendell Stanley as the chairman, an appointment Sproul maintained would "bring great distinction to the University." Predictably, the regents considered Sproul's case with deliberate speed—until the Nobel Prize committee announced that Stanley had won that year's most coveted prize for his groundbreaking work on viruses. Then with uncharacteristic swiftness, the regents offered Stanley a "grand salary of $9,600" and nearly complete authority to select his own staff and design and lead a new Biochemistry and Virus Laboratory (BVL).[13]

The decision to give Stanley control over the BVL seemed at the time an astute and reasonable decision. Many universities coveted this distinguished experimentalist for their own bioscience programs. Most agreed that with Stanley at the helm, the BVL would extend the line of success carried by Berkeley in the previous decades. "The whole enterprise that Dr. S(tanley) sketched," commented one observer, "is really a huge and perhaps Californian venture in developing biochemistry . . . there would be no question that the University of California will be the leader in [the biological sciences] five years from now." Another anticipated that with Stanley at the helm, "epoch-making discovery, more definite assumptions on the origins of life will be made than ever before." And the UC student newspaper, the *Daily Californian*, wondered how many Nobel Prizes would ultimately come to rest on the BVL's mantle.[14]

But in their haste to create a top-flight research program in the biological sciences, UC regents and administrators failed to recognize certain peculiar aspects of Stanley's selection that threatened the effective implementation of his vision. Success or failure of the BVL would hinge upon peculiar nonscientific issues. Could Stanley's elite-driven, autocratic managerial style work in the country's preeminent public university? Was Berkeley's intellectual base throughout the biological sciences too diverse to unify as a single research program? Would Berkeley's faculty accept Sproul's unilateral selection, or Stanley's near-dictatorial authority to oversee the entire laboratory? Moreover, would Stanley's decision to emphasize virus research, yet exclude staff conducting viral research in medicine and agriculture, antagonize the disciplinary divisions that already existed? Furthermore, virology has, by definition, obvious clinical and agricultural relevance; therefore, how would investigators in these fields interpret the inherent contradiction in Stanley's vision? Even if Stanley could wring order out of the chaos that plagued the biological sciences at Berkeley, it remained to be seen whether fundamental research was the best course of action.

To Stanley, however, the fate of his academic program had nothing to do with administrative matters, and even considered the highest quality of scientific research produced in his laboratory a foregone conclusion: "The first order of business," wrote Stanley in his letter accepting his appointment, "is to secure funding."[15]

Chapter 2

Patronage and Policy

Free scientists, following the free play of their imaginations, their curiosities, their hunches, their special prejudices, their undefended likes and dislikes. . . . One can no more produce fundamental and truly original work by means of some grand-over-all planning scheme for science than one can produce great sonnets by hiring poets by the hour.

—Warren Weaver

It is the fundamental tenet of our "religion" . . . that research must be free and researchers must be free.

—Robert Felix

Whether celebrated as saviors or scorned as meddlers, patrons of science play a major role in shaping research and influencing its pace and direction. Yet too much can be made of scientific patronage as a *cause* of discovery. Scientists do not always conduct their experiments in concert with the intent of their sponsor, and sometimes a single scientific discovery can destabilize entire fields. Patrons may provide much needed stability for an uncertain field, but their money nonetheless strengthens the entire community—laboratories, research equipment, staff, and experimentalists too.[1]

In the immediate aftermath of World War II, sponsors and scientists clashed over whose interests should hold more weight, establishing in effect a contest over intellectual claims and a struggle for disciplinary authority. Amid steadily rising commitments to scientific research in the postwar era, sponsors and scientists moved toward a full-scale collision. Nowhere was this collision more evident than in the biological sciences. Before World War II, the Rockefeller Foundation—the most powerful private sponsor of scientific research—cautiously guided a steady and methodical experimental field. Then, beginning in the late 1940s, the federal government implemented policy that directed unprecedented levels of money to support laboratory research, transforming such agencies as the National Institutes of Health from a once-insignificant government agency into a bureaucratic behemoth. In an instant, the state eclipsed the power of private sponsors.

Paradoxically, however, the powerful arrival of federal science policy launched the biological sciences on a path of increasing authority. This metamorphosis was clearly marked by the state's desire for institutional stability. Policymakers created an antiseptic infrastructure that gave scientists the authority to manage funding channels and staff peer-review boards. In effect, the federal policy that supported research in the biological sciences created a double transformation: the ascendancy of state-sponsored patronage and the capture of authority by biological scientists.[2]

Private Patronage

From all sides and at all times, unrelenting pressure plays upon academic scientists to locate sponsors willing to provide financial support for research. Private industry, an omnipresent source of potential patronage throughout the twentieth century, has always sought early access to experimental results, exclusive contact with graduates, or the right to establish dual licensing agreements on patentable discoveries. But some academic scientists believe that an apparently well-intentioned offer of corporate patronage conceals highly problematic trade-offs. To accept private money means entering a world from which an experimentalist may not return; to open that door means living forever with the messy, intractable problem of mission-oriented research. Investigators who choose to eschew corporate money may slog through years of perpetual impoverishment, but it is a condition much preferred to its antithesis—the loss of experimental autonomy or, worse, objectivity. The opinion that no condition, not even desperate financial need, can justify taking corporate money, would become in later decades one of the most conspicuous reversals of the field.

When the twentieth century opened, before there was aggressive sponsorship of the biological sciences, when industrial support still seemed antithetical to "objectivity," and in the absence of any formidable governmental mechanism in which to distribute patronage, most investigators survived on internal university support. American colleges had always been cloaked with a public purpose, with a responsibility to the past, present, and future. The creation of tax-supported universities through the Morrill Land Grant Acts of the nineteenth century, and expressions of Christian generosity that guided individual benefactors during the Progressive Era, advanced this charitable assumption further: the college was expected to give more than it received. But this was a less than predictable source of revenue for the biological sciences, a field that had shown more promise than production, more hope than substance. As a result, the first two decades of the twentieth century witnessed not only an effort to identify a scientific direction; they were also a trying time to locate financial support where little existed, of choosing experimental topics with little hope for lasting financial support. This perversely asymmetrical existence impeded any hope for immediate progress: to become productive, the biological sciences needed money, and to get money, the field needed to produce. Alongside this agonizing conundrum loomed a third obstacle: alternative revenue sources came with strings attached. To many biological scientists it seemed as though this experimental, professional, and fiscal nightmare might haunt the field forever.[3]

In the face of such a precarious financial existence and chaotic scientific state, biological scientists began turning toward private support offered by a small group of philanthropists. The undeniable pillar of this community was the Rockefeller Foundation. No scientist could afford to ignore the foundation's dazzling bankroll, but at least as important as its wealth was its mechanism of awarding support through "peer review," a unique administrative apparatus nurtured by John Rockefeller, who initially wanted greater input in the distribution of his own money. Out of this provincial review process emerged a powerful foundation led by a few administrative leaders and a handful of their closest and most trusted scientific colleagues. Rockefeller review boards would gather frequently to assess the quality of every proposal submitted and evaluate the skills and qualities of the program's principal investigators. They had broad discretionary powers to select project goals without disciplinary or site restrictions, other than a general regard for social good. The wide latitude given to Rockefeller review boards and their relaxed peer-review process sparked a hundredfold increase in foundation grants to universities for research during the 1920s; support ranged from the now-infamous eugenics studies to the underappreciated international farming projects that fed hundreds of thousands of people worldwide.[4]

Struggling mightily against a number of forces, in the 1930s biological scientists began experiencing another obstacle, in addition to the tightening of available funds during the Depression: a backlash of popular opinion. Haunted by unemployment, debt, insecurity, and the burning memory of World War I, the public began blaming scientists for helping corporations overproduce, militaries for murdering thousands of innocents, machines for making human labor obsolete, and manufacturers for making gadgets that few could afford. The instruments of traditional American progress—scientific and technological advance—had become for many the primary cause of prolonged human misery. Antiprogress groups sprang up, locally as well as nationally; Technocracy, the Committee on Technocracy, and the Technocrats preached a peculiar brand of social invention that contrasted with mechanical and scientific innovation. In reality, few people actually believed that something was wrong with science; however, many did wonder if scientists should have "authority on all social questions." As events would prove in later decades, similar public frustrations could reappear as a powerful counterforce to science, strike savagely and seemingly without warning, and needed to be taken seriously.[5]

It was in this tumultuous context that the Rockefeller Foundation launched a new deal for the biological sciences. In 1932, the foundation's leaders announced they would focus greater attention on the life sciences—nurturing, healing, socially responsive, and responsible medicine and biology—and promptly handed the reigns of this inchoate program to Warren Weaver, a classical mathematical physicist from old-school traditions. Weaver's entire life appears to follow that of the prototypical organization man. Born in the Midwest, Weaver grew up greatly influenced by his local church—a background from which he emerged with a set of largely conservative scientific convictions: a deeply rooted faith in the Protestant work ethic, a healthy respect for Presbyterian rationality and the scientific method, and the importance of socially purposeful science. His admirers were less interested in his intellect than in the measure of this man, and they saw in him a sincere and trustworthy leader. To them, he was the image of loyalty, fairness, and dignity; he was always guided by the highest ideals. Remarkably modest, Weaver gradually worked his way into academic administration, first at Wisconsin, then at CalTech, before heading off to lead the Rockefeller Foundation's new program to support research in the biological sciences.[6]

Weaver immediately set out with remarkable vigor to stimulate what the Rockefeller Foundation saw as an underdeveloped and disorganized field. He encouraged leading research universities such as CalTech and Chicago to become primary sites for research and earmarked millions of dollars in support to help them develop cooperative programs with physical scientists that might extend medical research beyond the domain of patient care. To distance his program from the foundation's embarrassing support of eugenics, race biology, and social hygiene in the previous decades, Weaver invoked with careful rhetoric a moral purpose to the new agenda. "The pursuit of an understanding of life," Weaver argued, "was just about the most moral activity that man can possibly devote himself to." "Knowledge is a 'good' in and of itself." To Weaver, the new mission for the Rockefeller Foundation was not a desire to control life, but the more dignified goal of understanding—notions compatible with pure research. No less important, the spirit to conduct research in the biological sciences, so underinspiring through the early decades of the twentieth century, had been infused with optimism and hope now that the Rockefeller Foundation would back the field.[7]

But for all the excitement that surrounded the new mission of the Rockefeller Foundation, it remained difficult to identify or define. Little coherent pattern could be detected in the unlikely mixture of directions that Weaver offered. His 1934 progress report to the foundation's board supported a wide range of research topics: "Can we obtain enough knowledge of physiological and psychobiology of sex so that man can bring this pervasive, highly important, and dangerous aspect of life under rational control? Can we unravel the tangled problems of endocrine glands . . . ? Can we solve the mysteries of vitamins . . . ? Can we release psychology from its present confusion and ineffectiveness . . . ?" One year later, Weaver contributed to the confusion with a series of articles for the *New York Times* that declared that the foundation would emphasize some vague notion of pure research. Later, he tried to simply avoid the indecision while he cultivated physicists to conduct research in biochemistry, cellular biology, and genetics.[8]

This was the confusing array of policy advice besetting biological scientists in their desperate search for patronage. Then, in the May 1938 issue of *Science*, Weaver offered a clearer portrait of his vision destined to become influential. Biologists must focus their studies on the "ultimate littleness of things," declared Weaver, which he described as "subcellular biology," or a "biology of molecules." Then he offered a simple phrase that would give immediate direction to the foundation's central purpose and serve as a guiding principle to which biologists could now aspire: "Molecular biology."[9]

In hindsight, Weaver's call for molecular biology established a coherent plan for experimental research in the biological sciences. Indeed, considering the litany of modest and somewhat ambiguous signals sent by the Rockefeller Foundation, Weaver's introduction of "molecular biology" provided direction—direction for an unfocused field, to be sure, and direction for investigators and university administrators to follow what many called the harder sciences, such as physics and chemistry, and to rely on these disciplines for much needed intellectual and technological leadership. Weaver did not try to compose a specific path for molecular biology; he merely wanted to narrow the base from which an independent discipline might emerge.

However, these histrionic celebrations should not obscure a central fact: the essential logic that informed Weaver's call for molecular biology was agonizingly ambiguous from the outset, and would remain so for many decades to come. Weaver had introduced molecular biology as a "long-term project" and expected researchers in the field to flounder about until some "fields of critical importance" provided leadership. Repeated conferences with biologists failed to produce a clearer understanding, so Weaver cut the Gordian knot by targeting a few experimental subfields and designating an even smaller number of American research universities as primary sites for development: the University of Chicago, CalTech, and Wisconsin. Wendell Stanley longed for the BVL at Berkeley to join this select group, but to secure support from the Rockefeller Foundation for some uncertain notion of molecular biology seemed a monumental task.[10]

Stanley may not have appreciated Weaver's support of other research universities, but he showed no shortage of interest either. This was the Rockefeller Foundation: the most powerful and generous private sponsor of biological inquiry in the world. Moreover, its sweeping influence and reliably generous support could set the stage for innumerable later ventures with other private sponsors. Submitting an application to the foundation was in this way an unavoidable necessity for Wendell Stanley and the University of California. But a number of questions persisted. What did the Rockefeller Foundation mean by *molecular* biology at the time, or, more precisely from Stanley's point of view, what criteria would the Rockefeller Foundation use to evaluate proposals for research in molecular biology when they had yet to define it themselves? Would scientific merit drive the review process, or would it be something else?

The Rockefeller Foundation and the BVL

In November 1947, Wendell Stanley submitted a proposal to the Rockefeller Foundation for "$500,000 to $600,000 to help cover construction costs, research support, equipment, operation and salaries" for a Biochemistry and Virus Laboratory at Berkeley. "It will become," he declared confidently, "the most powerful virus group in the World."[11]

By all accounts, Stanley's application was the strongest submitted to the Rockefeller Foundation in 1947, and perhaps stronger than anything that Weaver had seen throughout his tenure. Nor did anyone at the foundation deny that Stanley's grand plans had a very real chance of becoming reality. His proposal described powerful collaborative potential. It listed in great detail the technology the staff would use to ensure consistency in a new experimental field. And he expressed a sincere desire to take advantage of the "unique flanking strength" with Berkeley's physics department—without question, the strongest program in the postwar atomic age. Before Rockefeller officials could balk at an overemphasis on physics, Stanley also posed his lab as preeminently necessary for medical and agricultural research: "We should have proper facilities for work in case of a national emergency," to "defend" against either "the loss of life or crops to viruses." Finally, Stanley went to great lengths to describe the BVL in relation to "molecular biology," and he repeated the phrase in virtually every correspondence that he had with the foundation.[12]

Rockefeller officials admired Stanley's "primary objective . . . to elucidate the nature of virus reproduction," his commitment to "the basic biochemical and biophysics of . . . proteins," the willingness of the staff to rally around protein- and virus-centered research, the boldness in which he defended his vision, and the grand social impact that it promised. Overall, Stanley's proposal captured the attention of the Rockefeller Foundation like few others had, and it moved swiftly through the initial stages of the review process.[13]

On-site evaluations were equally sanguine: "[Stanley's] lab is clearly destined to be one of the largest and strongest centers in the world for research in biochemistry"; "at Berkeley, Stanley will put under one roof the best facilities for virus research to be found anywhere"; "there is no place in the world where a person can obtain better . . . instruction of laboratory research."[14] Rockefeller's on-site review committee immediately dashed off a letter sent directly to Warren Weaver, informing him in no uncertain terms that "there can be little doubt that this is a scientific proposal of greatest timeliness, a situation in an almost perfect setting, and with probably the best leadership available in the world. To that end [we] have in mind primarily the possibility of a grant of at least $1,000,000, and preferably larger."[15]

Warren Weaver, duly impressed, wanted to surprise Stanley and the Berkeley administration in person with an award that doubled their original request. Before he left on his trip to Berkeley, however, Weaver received a distressing letter from Stanley: "Looking over some of the reviews published by RF (i.e. one million to Lawrence, 6 million total to Cal Tech)," wrote Stanley, "it is obvious that the West Coast and especially the UC has not fared so well with respect to major grants. . . . To that end I would like to increase my original request to $1,000,000."[16]

Stanley's decision to accuse the Rockefeller Foundation of favoritism and his audacity to double his original request on such short notice—there is no evidence to suggest that he knew Weaver intended to personally deliver a $1-million grant—was a singular and eventually disastrous miscalculation. In addition to leaning on a profoundly illogical argument to support his accusation that the Rockefeller Foundation favored East Coast research—CalTech and Berkeley were of course on the West Coast—Stanley's lack of social graces and his nerve to request a larger gift shattered the mask of decorum and formality that had implicitly become a crucial part of the Rockefeller's application process, especially considering that the criteria for evaluating the scientific merits of research in molecular biology remained so elusive.

Word spread like wildfire throughout the Rockefeller community that Stanley had offended Warren Weaver: "'W' (Weaver) emphatically dislikes the fact that California [UC administrators] and Stanley thinks Stanley has all sorts of important connections which will assure large support for him." Weaver immediately postponed his trip West to reflect on the decision to double Stanley's request, and he ordered the foundation to freeze Stanley's award money because "virus research has not reached a stage at which substantial investment is likely to bring substantial results"—an impulsive decision that contradicted years of dogged Rockefeller support for virus research, including Stanley's extraordinary work on the tobacco-mosaic virus at Princeton. In a private letter, Weaver admitted that he "dislike[d] Stanley's manner of approach so much that it is hard to think clearly about the proposal itself . . . I frankly don't know if Stanley is the best person to direct such a many faceted project."[17]

Any hope that Stanley had in mending the strained relationship quickly disappeared when he inexplicably sent off another antagonistic letter that again challenged the Rockefeller Foundation bias for the East Coast. Barely able to contain his anger, Weaver wrote to Stanley:

> [W]e have, in the 16 years since I joined the staff, devoted a total aid within the state of California which happens to be greater than the corresponding total for any other state in the union. . . . We have not at all attempted to level off the development in various areas . . . in terms of geography, but in terms of opportunity. To consider whether or not we have given similar aid to different states would be from our point of view be [*sic*] like complaining to the mining industry because it does not mine the same amount of gold in every state.[18]

Despite justifiable misgivings, a few administrators at Rockefeller nevertheless remained enamored with Stanley's proposal and the promise that it held for future molecular biology research. But Weaver, at this point, refused to back down. To him, Stanley was "a very able, but very cocky individual." "He will be shocked to learn," wrote Weaver, "that this [has become] a rather charged situation for us. I favor giving Stanley $150,000 for equipment. I do not favor giving him what he has most recently requested—$1,000,000."[19]

In overestimating his own reputation and underestimating the Rockefeller Foundation's pride, Stanley sabotaged any opportunity to secure a substantial amount of Rockefeller money for the BVL. Yet he continued to pressure the foundation for more funds in a series of letters almost too painful to read: "I think it would be more sensible to now ask for . . . as much as $350,000 . . . for special equipment . . . needed immediately."

Completely oblivious to the fact that he had, on this occasion and before, offended the largest and most influential patron of the biological sciences in the world, Stanley closed the letter by reminding Weaver that the BVL "exemplifies in the highest degree, a new program for 'molecular biology.'"[20]

Stanley's letter may have positioned the BVL to become the Rockefeller Foundation's favorite son for research in molecular biology, but Weaver responded to the latest and subsequent requests with notes that said simply, "No." Still, Stanley continued to press for money. To his request for "$250,000/3 years" Weaver replied, "there is simply no chance at this moment to consider a different grant." Stanley kept up his barrage until Weaver effectively ended their relationship when he reduced again the total amount of support that the Rockefeller Foundation would award the University of California's BVL, from $150,000 to $100,000. Weaver would later write in his diary a prescient observation of his yearlong nightmare with Stanley:

> [I]t impresses [me] more and more that there are some scientists who, as all-around scholars and as human beings, turn out to be vastly greater than any of their greatest discoveries. [These] great men train and inspire many excellent young persons, and have a profound intellectual and moral influence upon their whole surroundings. On the other hand, there are always the somewhat pathetic instances of scientists who do not turn out to be as great as their greatest discovery. [I] would have to put Stanley in this second category. His original work on the crystallization of the virus of the mosaic disease of tobacco was undoubtedly a great discovery; but [I] simply cannot convince myself that S (Stanley) is truly a great man.[21]

The Rockefeller Foundation helped establish the basis for a new and potentially lasting direction for the biological sciences. In detailed examinations of Rockefeller funding received by CalTech, Chicago, and the University of Wisconsin at Madison, it is clear that the foundation leaned toward any proposal that took to a "molecular vision of life," emphasized large-scale, technology-centered, fundamental research on proteins and viruses, and downplayed or avoided altogether medical concerns. Initially, Stanley's application satisfied these tendencies. Though Weaver may not have been clear about the criteria his foundation used to evaluate proposals for molecular biology, Stanley's decision to merge the physical and life sciences and study the "structure and specificity of virus molecules" certainly jibed with Weaver's desire to promote in the mainstream an almost entirely new discipline. Against this backdrop, what stands out about the scientific basis of Stanley's proposal is that Weaver thought he saw a "bold vision of consequent possibilities."[22]

Just as clear, the Rockefeller Foundation decided against Stanley's

BVL for decidedly nonscientific reasons. At times, Weaver and his cohorts emphasized physical and chemical studies; on other occasions, they demanded greater attention to the relationship between biological matter and human behavior. Furthermore, or perhaps because of these ambiguities, the Rockefeller Foundation determined that molecular biology would emerge out of a large-scale project led by an established scholar in a traditional research program, such as at CalTech, Chicago, and Madison. Or perhaps because they lacked sharply defined scientific standards, Rockefeller officials had little choice but to place greater emphasis on personal relationships as criteria in which to measure the worthiness of the BVL. Given such sensitivities, Wendell Stanley could plausibly claim that the Rockefeller Foundation's decision spelled a messy tendency of private money—then and later—to occasionally prefer nonscientific concerns over truly innovative scientific research projects.[23]

A Federal Policy for Bioscience Research

As the 1940s came to a close, neither Berkeley, Stanford, nor UCSF had a notable research program in the biological sciences. Without a solid foundation and facing an uncertain future, biology at these three universities slept while the physical sciences and engineering struggled mightily with expanding faculties and explosive student enrollments, the chaos exemplified by the Quonset hut–like "portables" that were still scattered about the campuses and by the tents that served as temporary classrooms until more permanent facilities could be built. No one at these three universities could point to anything remotely exceptional in the biological sciences, or could envision the vibrant center of research that federal policy was to make.

Nor could anyone have foreseen that the Russians would provide the spark that would launch the biological sciences into a new age. Almost immediately after the war, Communism showed itself to have much greater influence around the world than previously imagined: among some of the more notable examples, in 1948, the Soviets clamped a blockade on East Berlin; in early 1949 they detonated their first atomic bomb; then, months later, American diplomats declared that China had "fallen" to the communists under Mao Tse-tung. Events such as these confirmed for many policy officials that Communism posed a clear and present danger to the free world, threw down the gauntlet to passivity, and heralded a crusade to boost America's entire scientific community. As doubts grew about Soviet intentions abroad, concern about scientific strength at home became obsessive. But when policy officials talked about reviving the sciences, they usually spoke about research to improve missiles or develop new weapons—generally treating the biological sciences as an afterthought.[24]

Haunted by the fear of international tension, scientific underperformance, and nuclear war, many policymakers began to think about scientific development as the nation's principal concern. Congress picked up the debate that had begun during the war over the proper direction of postwar science and the proper relationship between scientists and the government. Throughout prolonged committee hearings, elected officials spoke openly about creating policy that would tie research more closely to public welfare and narrowing the distance that separated experts from democratic control. Various legislators drafted bills that espoused a vision of science policy that fit within the framework of the New Deal political economy. Most drafts were vague and inchoate, but they offered a creative and genuine attempt to wrestle with difficult questions about the social role of science and its place within the evolving political economy of postwar America. As policymakers began drafting bills, criticism sprung from a variety of corners. Isolationists, opponents of military funding, anti–New Dealers, and a hawkish faction that nevertheless loathed open scientific exchange emerged as powerful counterforce to science policy debates. Ironically, scientists held the keys to cloture, a decisive political influence that the next generation of scientists would not possess.[25]

The brute problem for policymakers was the anxiety that scientists felt about the federal government's interest in their work. Scientists may have accepted their relationship with their government during the war, but federal policy for research during peacetime was an altogether different matter. Biological scientists proved especially vocal in their contempt. Hubert Loring, a biochemist at Stanford University, firmly believed that patronage from any external source threatened scientific objectivity, and found it difficult to accept the idea that the federal government had a "benign" interest in his work. "Several of us at Stanford," warned a colleague, "are afraid that it is becoming more and more obvious that the only way we are going to be able to continue scientific work is to turn our efforts more and more toward applied lines." Another worried that while the military had once been "unduly slow in some cases to take up new ideas developed by civilian scientists . . . now, in the wake of the bomb and the Cold War . . . the military, if anything, has become vastly too much impressed with the abilities of research and development." Private patrons of scientific research, such as Warren Weaver, also doubted "whether it is either possible or desirable to carry over into peacetime research, many of the elements of organization and control which properly and inevitably characterize wartime work." However, the primary obstacle that beset policymakers remained scientists' unrelenting demand for autonomy—they used the word "freedom" often to emphasize their indispensable desire to follow their own research interests, even if their work had no obvious purpose.[26]

Historically, the relationship between scientists and the state had been growing for quite some time. In the late nineteenth century, the Department of Agriculture sponsored research designed to help the nation's farmers improve their productivity. That relationship continued to blossom throughout the Progressive Era, and then, with the onset of World War II, exploded into a bewildering host of new science research support agencies. The War Production Board, with the power to compel private industry to address scientific and technological needs, replaced the tepid National Defense Mediation Board in January 1942. The War Manpower Commission appeared a few months later, charged in part with directing scientific research toward national and military needs. At about the same time, the Office of Scientific Research and Development (OSRD) established, among other projects, the Committee on Medical Research in order to coordinate the efforts of universities and pharmaceutical companies to increase research and production of antibiotics.[27]

The federal government's extremely generous reimbursement policy during World War II also won over some converts. The OSRD offered "contract overhead," which promised to remit scientists for up to 50 percent of all their costs to encourage research considered relevant for the military. As a further emolument, other agencies reimbursed scientists on a cost-plus basis, providing iron-clad guarantees that they would cover all research expenses. Furthermore, to encourage investigators to "voluntarily" accept federal money, sponsoring agencies relaxed overhead guidelines and accepted virtually any expense affiliated with work on war-related projects. Though critics found many instances in which individual researchers had abused the government's overly generous overhead policy—a charge that proponents did not deny—many federal officials accepted "abuse" as an unavoidable tax during wartime emergency. Although the OSRD was dismantled at the end of the war, "contract overhead"—known informally within university circles as "soft money"—became the basis of science policy during peacetime.[28]

No less necessary to the extension of national science agency beyond wartime, scientists required a strong foothold within government sponsoring agencies that would allow them to protect their cherished autonomy. To win their approval, some policymakers offered provisions that duplicated the peer-review process popularized by the Rockefeller Foundation in the previous decade. But before policymakers wrote a scientist-friendly bill, a freshman senator from West Virginia, New Dealer Harvey Kilgore, submitted S.R. 702—known as the Science Mobilization bill—that placed a federal science agency under the direction of the President. Most scientists vehemently opposed Kilgore's bill because they rightly expected that national science policy placed within the reach of the Oval Office would inevitably become a political campaign issue hostage to the fickle winds of popular opinion. And, for the scientists at least, that was completely untenable.[29]

Research scientists could hardly have asked for a more effective political point man to defeat Kilgore's bill than Vannevar Bush, scientist, engineer and director of the OSRD. A political conservative and certainly not a supporter of the New Deal, Bush had earned the respect of the academy when he condemned Kilgore's amendment as "hopelessly political," and the respect of Congress with his skillful handling of the OSRD during World War II. Scientific productivity and heavy-handed federal policy were mutually exclusive, said Bush, and he hit this point repeatedly. Twisting the principles of Jeffersonian liberty beyond recognition, Bush acidly reminded legislators at a special congressional session that "our Constitution is a political instrument . . . [which] guarantees democracy and freedom, in which [all] people . . . decide what they want"—including, apparently, the protection of every investigator's right to direct their own research. When pressed for details, Bush insisted that a national science policy was responsible, but to be effective, it should not be responsive to anyone, not the president, not the legislature, and most of all, certainly not the public. He told federal officials that not all research contributed to stronger national defense; "basic research" was the domain of the scientist and needed to be protected; "applied research" was the domain of engineers, private industry, and political debates, and was influenced by popular opinion or commercial markets. Write policy that would allow scientists to run free with pure research, said Bush, and practical applications would naturally follow.[30]

Bush's stony civics lesson to Congress may have offended those who controlled the purse strings, but he galvanized scientists because he championed their professional autonomy. In the words of one scientific staff member, Bush's proposal was "an instant smash hit" within the scientific community. And as for the need of a federal agency permanently committed to supporting long-term basic research, Bush minced no words: "under the pressure for immediate results," Bush wrote in his influential treatise *Science: The Endless Frontier*, "applied research invariably drives out pure . . . unless deliberate policies are set up to guard against this." In this single text, Bush accurately foretold the greater part of the concern that scientists had toward the sponsorship of their research by the federal government during the postwar period.[31]

In March 1950, after nearly six years of false starts, re-drafts, and political missteps, Congress finally succumbed to the will of science and passed the bill that created the National Science Foundation (NSF). The final bill leaned significantly toward the version supported by Vannevar Bush: the government would be the "chief sponsor of basic research" but would place scientists in critical administrative and advisory positions. The only thing that scientists did not get with the NSF was a laissez-faire organizational structure that lacked enforcement powers, but even this compromise was, paradoxically, a victory for scientific autonomy and pure research. In theory, new science agencies like the NSF were commanding super-agencies that extended from and ensured the execution of popular will. But in practice, political representatives charged with overseeing these agencies were far less formidable than they appeared when drafting the bills in committee. Federal representatives rarely understood the details or complexities of the cutting-edge scientific research they were charged with evaluating. Even more telling, individual investigators advising political representatives—or more often, their aides—had difficulty explaining scientific intricacies in a way that the nonscientist could understand, especially the more esoteric projects dedicated to fundamental research. Senator Kilgore, a staunch proponent of a national science policy, nevertheless confessed his "utter, absolute ignorance of science."[32]

It was partly out of necessity, then, that scientists rather than politicians controlled national science agencies such as the NSF. For instance, Congress allowed scientists to duplicate in the NSF and other federal agencies the Rockefeller Foundation–like "relaxed peer-review" process to evaluate requests for funding, which effectively handed over the reins of these public agencies to the scientists themselves. Voluntary peer-review boards, comprised of eminent scientists from universities, hospitals, foundations, and other research institutes, quickly became the standard by which the NSF and almost all other federal agencies evaluated grant proposals and distributed money. According to one scientist who volunteered as a peer reviewer, the trend of committees like his was to fund all research projects, "so long as there was some relation to or possibility of a new discovery."[33]

This is not to say that the federal government stayed out of scientific laboratories. On the contrary, loyalty oaths and background checks were only the most obvious example of the government's attempt to check the autonomy of the scientific community. Indeed, before the decade was out, the State Department suspended or revoked the passports of nearly 40 leading investigators, including the Nobel laureate Linus Pauling. However, peer-review boards in federal agencies such as the NSF sidestepped the messy controversy of loyalty by granting award money to the applicant's university for distribution, which effectively made university administrators responsible for the award, background checks, and oversight. Ultimately, federal officials committed to fighting the cold war groped for ways to mobilize scientists without obstructing their research.[34]

From this hurried, chaotic, initially defensive, and ultimately compromised "national pressure" for greater scientific output came policy that generated an explosion of federal programs. But the creation of a more productive scientific community was not simply a matter of asking researchers to spend more time in the laboratory. Buying new equipment, hiring new investigators, attracting and training students, and, above all, diverting energy and resources from other fields all proved enormously expensive. But Congress always complied. By 1950, federal expenditures on scientific research were growing at an average rate of 14 percent per year, more than three times the growth rate of the country's gross national product. Both houses of Congress established standing committees on science and space and immediately began budgeting almost $1 billion annually to promote "advancing scientific knowledge," with the Office of Naval Research, Department of Defense, and the Atomic Energy Commission (AEC) receiving almost one-half of that total. The biological sciences benefited from the spillover, where financial support for research increased from a stingy $180,000 distributed through approximately fifty contracts to a much more respectable 264 contracts that offered a hefty $4 million in support.[35]

And it would get even better for the biological sciences after the outbreak of the Korean War.[36]

No agency better exemplified the jostling and frenzied expansion of federal science policy than the NIH. Wilbur Cohen, future secretary of Health, Education and Welfare, considered the NIH a "brilliant jewel." Others, including Representative Gordon Canfield of New Jersey, described it as "America at its best." By whatever description, the NIH underwent a profound transformation during the early cold war years from a modest biomedical unit into the most powerful biomedical research center in the world. The origins of the NIH can be traced back to 1887, when Congress budgeted just under $50,000 for a small agency called the Public Health Service (PHS), and stipulated that certain infectious diseases, such as cholera, smallpox, and diphtheria, would receive the most attention. Then, during the early Depression years, the rising threat of environmental and so-called chronic diseases—especially cancer and heart disease—prompted Congress to spin off the NIH from the PHS, and then provided the adolescent agency with a total budget of $43,000. Three years later, Congress further tied medical research to social welfare when it wrote into the Social Security Act of 1935 the Title VI clause that assigned the NIH primary responsibility for leading an "investigation of disease and problems of sanitation" and increased its budget to $375,000. Three years after that, Congress established under the direction of the NIH the National Cancer Institute (NCI) in order to promote a concentrated attack on cancer and added $300,000 to the NIH budget. To improve the treatment of infectious diseases such as typhus and promote the development of blood plasmas, fluoridation, and the penicillins during World War II, the OSRD went into high gear: it shifted scientific personnel into biomedical research; increased funding of off-site extramural projects at independent universities, laboratories, and institutes; took over operation of the medical research arm of the NIH; and spun off the Centers for Disease Control. By 1945, successive national crises in the 1930s and early 1940s made biomedical research critically important and at the same time turned a smallish NIH into what one scholar describes as a behemoth and made "applied public health measures" of "secondary" importance.[37]

The size of the NIH may have increased substantially during the war, but in the early postwar period, Congress considered its "frugal budgets" and "limited scope" in thorough need of revival. For instance, Democratic Senator Lister Hill reminded his colleagues: "The development of a solid foundation of fundamental knowledge concerning biological processes is an essential condition for breakthroughs." Republican Representative Ben Franklin Jensen from Iowa, who regularly offered appropriation amendments that cut all budget requests by 10 percent in every category, consistently exempted the NIH from such economy. To the director of the NIH, Representative Jensen intoned: "I hope, Doctor, that if you need more funds, you will ask for them. This is one place where [we are] quite liberal. I do not think we should worry about a few million dollars when it comes to finding the cause and cure of these dread diseases. To do other than what this [Congress] has been doing in furnishing these funds, I think would be almost to the point of criminal."[38]

Senator Robert Byrd considered recent appropriations to the NIH as "infinitesimal" and demanded immediate increases in all future budgets. Senator Henry Dvorak, whose natural tendency was to save money, and whose work on budget committees made him more than occasionally nervous, nevertheless got caught up in the mystique of the biological sciences when he asked one NIH official if he had enough "appropriations to attain [his] ultimate objectives." Representative Walter Judd from Minnesota, who as a physician had special insight into the biological sciences and who had originally doubted the value of national science agencies, reversed his original opinion and declared that "the National Institutes of Health should have no peer anywhere in the world." Sometimes Congress' sweeping directive to push biology's frontier staked out a position well beyond what individual investigators were prepared for, even getting a little belligerent with any investigator who appeared reluctant to take the public's money:

SENATOR HILL:	What about your . . . facilities. . . . Do you have sufficient space?
WITNESS:	Operations no longer require more electronic equipment. For instance . . .
SENATOR SMITH:	Did you say you had sufficient facilities?
WITNESS:	I have not said yet. . . .
SENATOR HILL:	This is what I want to know: *how much money do you want?*

From fiscal conservatives to New Deal liberals, few opposed spending public money on biological sciences, and when they did object, as in the case of Senator William Proxmire of Wisconsin, they found the pressure to fund the NIH so great that they "couldn't do anything but go along with the increases." According to one observer, bioscience research had become in the early cold war years "almost as much of an obsession to Congress as it was for scientists at the NIH."[39]

Straining to meet the ambitious goals and expectations Congress had set, NIH review boards loosened their imaginations, abandoned any vestige of discipline, and lost sight of balance between relevant and fundamental research. In the first two years after the war, half of the NIH's $10-million budget supported extramural research; then, between 1948 and 1954, total NIH budget appropriations surged beyond $300 million, with almost 70 percent supporting independent research projects. In 1946, the PHS granted forty-four extramural research contracts; by 1950, the NIH awarded 1,115 extramural grants, for a total of $11,508,841. Research fellowships for graduate students also grew from "a few tens of thousands of dollars" in 1948 to somewhere between $600,000 and $700,000 by 1952. Sometimes it was difficult to keep track of the change: in 1946, the number of NIH study sections increased from zero to twenty-one; the next year NIH officials estimated that around 250 to 300 study sections existed. In 1950, the National Cancer Institute awarded sixty-three grants totaling more than $16 million in its first year of operation. With the Korean conflict hovering in the background, the NIH added five more institutes—microbiology, arthritis, metabolism, neurology, and blindness joining the list—granted $16,374,128 to support extramural research on an operating budget of $60 million, and pluralized its name to "Institutes."[40]

The history of the NIH merely captures in compressed form the grand explosion of federal agencies, divisions, and institutes that supported the biological sciences during the early days of the cold war. Many observers of national science policy generally admire the patience and generosity of the federal government to sponsor research in the postwar period; in reality, the incredible surge of investment in the biological sciences was more a bureaucratic frenzy than shrewd calculation. As one science advisor noted at the time, "in reality, there was really no such thing as government research policy." Indeed, in the mass of jerry-built agencies there were pieces of policy—in the form of instructions covering procedural matters, for instance—but no policy in the sense of a coherent, local body of principles clearly stated and openly followed. Rather, federal policy for the biological sciences spun out, at first as an afterthought, then on a piecemeal basis, and finally as the summation of an extraordinary commitment.[41]

The complex arrangements, assortment of divisions, multiple organizational levels, and mixture of federal departments diverted bureaucratic attention away from scientific concerns and toward institutional territoriality in which funding agencies often took great pains to know about the research projects of other departments, or spending money on projects that might one day end up as part of someone else's program. Thus, while specific agencies may have assumed primary responsibility for overseeing a specific type of research topic, they often displayed a great willingness to fund virtually any project that came their way, especially those that lay categorically outside their jurisdictional boundaries. Consider that in addition to the NIH, a significant proportion of research funding also came from the Pentagon, hardly a unified voice itself, as the services typically fought fiercely among themselves. Within the Office of Naval Research, for example, one-third of its budget was spent on pure life-science research projects remotely related to naval purposes—including the study of animal behavior, underwater acoustics, and communication among dolphins and other marine animals—but it also supported a wide range of alternative projects in the biological sciences, such as crystallography research. Such interagency warfare at the federal level, certainly inefficient as a bureaucratic procedure, nevertheless set the stage for an explosive increase in money available for bioscience research.[42]

The Policy of Prosperity for Stanley and the BVL

As federal policy trends became clear, a sense of high excitement gripped Stanley, faculty in the BVL, and the UC administration. As eager for federal support of research as they were resolute in their rejection of federal intervention in their affairs, Stanley and UC administrators abandoned any vestiges of control and sought as much government money as possible. Measured against the serious reservations that the

Rockefeller Foundation had toward Stanley and the BVL, the relaxed review process of federal science policy must have been a welcome relief. In the inescapable trade-off between quality and quantity, the Rockefeller Foundation characteristically chose the former, the federal government the latter. Rockefeller officials tirelessly pored over grant applications in search of "qualitative superiority." Federal policy that supported research in the biological sciences, by contrast, had so many participating agencies that their collective decisions promoted variety and, in some cases, even relaxed the demand for quality in order to promote greater scientific output. Consequently, while the Rockefeller Foundation and other private foundations determined that Stanley deserved only partial funding, the AEC, ONR, and NIH showed an increasing willingness to provide funding for the construction of a physical plant, experimental equipment, and research and graduate training. Indeed, the BVL received a total of seven separate grants during the 1949–50 academic year, and Stanley still had not yet moved into his office.[43]

Less known but perhaps more significant money came from an unlikely source: the California state legislature. Since the University of California was a publicly funded institution, Stanley found it easy to play to the issues that mattered most to the voters and local politicians: cancer, education, and the cold war. Through the late 1940s, California legislators took turns pointing to new crises so grave that only state dollars could rescue them: schools were underperforming, Russian scientists were outproducing, and deaths to cancer were overwhelming, and the passing of more than one state assemblyman to this dread disease served as a constant reminder that medical research was a vast and uncharted scientific territory. Wendell Stanley unashamedly used these popular concerns—cancer, education, and the cold war—as tactical angles in which to sell the BVL. To those who saw education as an agent for change, Stanley celebrated the state's unwavering support of educational issues in the past and championed the need to continue funding instructional programs like the BVL. To the plethora of UC alumni in the state legislature, Stanley played on Berkeley's elite academic reputation and its contributions to the expanding California economy. To those concerned about public health, he promised to develop a better cancer treatment program. Any practical outcome could be made to look as if it fell within the scope of the BVL; indeed, virus research made his laboratory perfectly suited to "find the cause of such diseases as measles, mumps, chicken pox, influenza, and polio . . . and aid . . . agriculture by studying viruses which attack citrus trees, field crops, and farm animals." Finally, to those who feared the advancing threat of the Soviet technocracy, he promised incredible scientific advances; President

Sproul and Wendell Stanley compared the potential of the BVL with the contributions of Ernest Lawrence's cyclotron to the development of atomic weapons. It should be noted that Stanley and Berkeley's public relations department may have used popular issues to drum up support, but in private correspondence, they clearly intended to protect their professional autonomy.[44]

Unaware of Stanley's underlying intentions, California legislators responded with the most ambitious effort ever made by a state political body to furnish in-kind support for the biological sciences. Climaxing just one month of debates, in 1948, the state legislature passed unanimously Bill 569, known as "the Cancer Bill," which allocated $1 million of public money to help pay for the construction of biomedical research facilities across the state. Moreover, it included an additional $250,000 above the original allocation to direct "research on the origin, prevention and cure of cancer." Weeks later, state representatives decided to make the BVL the state's ubiquitous centerpiece in their new campaign to build the biological sciences and earmarked $600,000 specifically for the BVL, then increased that amount to $1 million, and then rewrote it again with a dramatically more generous appropriation—$1,215,000—with an additional $500,000 for unanticipated research expenses. Fittingly, state officials "urgently" agreed to cover all other expenses one year later when costs for the construction of the BVL soared past original estimates to more than $1.7 million.[45]

In Stanley's mind, the incessant need for experimental autonomy meant protecting individual investigators from federal interference, not rejecting federal aid to their programs. Such ideological agility was indispensable to investigators and university administrators such as Stanley who were intent on reaping the benefits of federally financed research while maintaining the experimental autonomy they traditionally enjoyed. After years of criticizing the intrusion of outside influences on their laboratories, investigators and university administrators considered patronage from federal programs a defining feature and an unfettered right of their field. A confused experimentalist boasted that recent "annual sales" are $700,000: "I haven't made a profit, but I haven't had a loss, either." A young Clark Kerr, soon-to-be chancellor of Berkeley, would make federal support of Berkeley's large laboratories staffed with nonfaculty researchers and professors engaged in "managing contracts and projects, guiding teams and assistants, bossing crews of technicians" the critical feature of his "multiversity." After a long, acrimonious standoff with sponsors seeking to intrude upon his laboratory, Wendell Stanley could complain in the early 1950s that government science was "ill-funded" and needed to have its appropriations increased "ten times their present level."[46]

That policy officials saw a relationship between popular concerns—educating children, national security, and curing cancer—and the BVL symbolizes this period as a golden age for the biological sciences. Even though the laws of limited resources clearly state that a disproportionate commitment to a narrow scientific topic can work at crosspurposes, more than one Bay Area newspaper celebrated unhesitatingly that "history's most concentrated attack on virus diseases—one of man's deadliest enemies—will get underway . . . on the Berkeley campus." Another applauded the courage of the BVL to "wage war on cancer." Sometimes the public would even speak of applying the "successes" of the Manhattan Project to public health without ever taking time to note the ironic contrast between weapons development and medical care. Though the goals of individual investigators and the public were often incompatible, the financial and popular support the biological sciences received in the late 1940s and early 1950s moved in the same direction, namely, the building of remarkably productive facilities with public money that were responsive only rhetorically, but not in practice, to public concerns.[47]

Federal science policy was not by itself a turning point for the biological sciences in the Bay Area. That had already occurred in early 1946, when Chancellor Sproul embarked, however uncertain, on a mission to establish the biological sciences at Berkeley, and then recruited Wendell Stanley to lead that charge. Still, the commitment of the federal government to making policy that would support scientific research was a highly significant moment of a sort that biological scientists had not previously experienced. The government's sweeping commitment, providing funding for virtually any experimental project, established a powerful foundation and set the stage for phenomenal later growth that in turn led to even greater expansion of the field. Federal policy was, in this way, a major step. Equally important was the autonomy that the profession retained—and in some sense, expanded—with the aid of federal patronage. The immediate lesson was that biological scientists did not actually need to worry about the centralized power of the state, or fear that an aggressive policy would encroach upon their professional realm. At the helm of the existing federal agencies, biological scientists could be certain that there would be more than enough money, no less autonomy, and no extraordinary pressure to conduct relevant research—just so long as the public was patient with the isolation of the science they had funded.

Despite its suspect organizational format, federal science policy initiated a historic drive that would invigorate the biological sciences in the Bay Area, as elsewhere. State money and scientists' virtual control of it would rearrange scientific disciplines and many sacred scientific assumptions. The haphazard funding process actually contributed to scientific serendipity, chance discovery, and if not by design, then by accident, the strengthening of an entire field. Despite the occasionally bitter tension between investigators and state representatives, the biological sciences as a whole benefited significantly, and disproportionately. Ironically, however, in the midst of expanding federal policy and rapid institutional growth, biological scientists failed to recognize the inherent tensions building between competing subdisciplines, or that the tactics used to construct their awesome cathedrals of research would one day hasten the turn of public opinion against them.

For Wendell Stanley, new federal policy for science brought more money for the BVL than he had ever dreamed. The biological sciences were bursting with ideas; times were good, and the future promised to be better. He was the director of what was destined to become a productive research laboratory, perhaps the best. He started to think about how to organize research, whom to hire, and what experiments to conduct. The BVL was the first to capitalize on new federal policy in this new scientific era.

And it would also be the first to fall.

Chapter 3

The Promise and Peril of the BVL

[Stanley] is actually designing the whole Department of Biochemistry to provide support for his virus work, all his new appointments having been made with this in mind.

—Frank K. Loomis

Since the trend was to gripe, I spoke strongly about . . . the one-sided policy of Stanley.

—Paul Stumpf

When Wendell Stanley arrived at Berkeley in 1948 as head of the biochemistry department and Virus Laboratory, his new colleagues welcomed him by urging everyone—staff, students, and the community—to "offer their complete support." But Stanley did not need any more accoutrements of power to direct a high-powered research project; he had everything he needed. Federal and state science agencies had awarded millions of dollars in support, and the Berkeley administration had given him absolute authority to make all staffing decisions and the option to select "position and salaries that should be offered" to any personnel who he wanted on board.[1]

But federal and state patronage provided something more than just financial security or professional authority for Stanley; it also delivered scientific peace of mind. It guaranteed the permanence of the BVL. More to the point, it guaranteed that Stanley's appointments and their research would last well beyond the short term. No one under Stanley should feel the pressure to discover something out of the ordinary to attract research support; everyone could work comfortably and carry on as they wished. In theory, faculty in the biochemistry department and at the BVL could afford to take scientific risks, while underfunded programs would set a cowering and unadventurous course in search of safe scientific haven.

No sooner had Stanley arrived at Berkeley than his dream for the BVL threatened to degenerate into a nightmare. Most ominous, a few unsubstantiated theories about DNA challenged the scientific assumptions upon which the BVL rested. Less worrisome—or so Stanley thought—parochial professional rivalries ran deep, and more dangerously, provided a constant and unwanted source of pressure against his authority. Stanley had no interest in rethinking his purpose, and so he organized scientific research and managed social relationships to cordon the BVL from unwanted external pressures. Stanley never made a conscious decision to run the BVL in this way; rather, the BVL was part of an ever-expanding idea—that pure research was superior to applied—that was drawing its strength from increasing federal and state funding.

The story of Stanley and the BVL is unusual, but the entire episode is not unique. The famous conflict that privileged pure research over applied at Berkeley was repeated countless times in laboratories all over, even serving as the framework that would soon organize the biological sciences at Stanford and UCSF, and it stands in stark contrast to the making of biotechnology in the next generation.

The Rise of the BVL

Berkeley's new biochemistry department and Virus Laboratory surfaced at a stable scientific moment. Most biological scientists believed that protein was the substance that made up all viruses, hormones, antibodies, and anything else that the human body recognized as disease. No one contributed more to the singular importance of the "protein paradigm" than Wendell Stanley, and not without cause. He won the Nobel Prize because he had supposedly found incontrovertible proof that a virus was in theory, if not in fact, pure protein. Stanley then took to spreading the gospel of the protein paradigm by freely distributing purified and crystallized tobacco-mosaic virus to university and private research laboratories. By extension, Stanley's disciples often dismissed deoxyribonucleic acid (DNA) as a useless "stupid" molecule, a skeletal frame from which hung the more prominent protein.[2]

Stanley and his minions did not dream this dream alone. *Scientific American* considered Stanley's identification of protein in virus as the final proof of the biological substance that "determined" life. Extolling the virtues of its native son, officials at the Rockefeller Institute raved that Stanley's discovery showed that "all that . . . we term 'life' is . . . made up to a very large extent of proteins." One of Stanley's lab partners at Princeton stated unequivocally that with Stanley's experiment, "we now know that proteins enter into . . . every vital process. They are the principal component, . . . the basic building stuff . . . of each cell of every living thing." Even the most revered scientists, such as Linus Pauling, used the protein paradigm to build remarkably productive careers—he was one of two people to win two Nobel Prizes—despite later acknowledging that he had "little direct evidence supporting basic assumptions of the . . . self-reproducing nature of proteins."[3]

Meanwhile, to the astonishment of protein's most hardened advocates, a few reluctant investigators were beginning to entertain the extravagantly far-reaching idea that DNA, rather than protein, might in fact hold all of the information for heredity in every living thing. As early as 1944, Oswald Avery found that he could remove all protein from a virus, but he could not rid it entirely of its nucleic acid. Two years later, a twenty-two-year-old Ph.D. candidate in genetics, Joshua Lederberg, announced that he had exchanged genetic markers between organisms without the presence of any protein. And two years after that, in 1948, Alfred Hershey declared that DNA was actually comprised of four unevenly distributed bases—adenine, thymine, guanine, and cytosine—which suggested that DNA was anything but a large, "stupid" molecule.[4]

The spread of DNA research had a harmful consequence for Stanley's professional stature, but hardly a fatal one. "I don't see much future in DNA," Stanley crowed with confidence to his cronies. Most contemporaries agreed. Prevailing wisdom dismissed Avery's article, probably because it appeared in a clinical publication, the *Journal of Experimental Medicine*, and his awkward, reluctant observation—"it is, of course, quite possible that the biological activity of the . . . nucleic acid is due to minute amounts of some other . . . undetermined substance [proteins?]"—provided enough of an opening for contemporaries to doubt his full hypothesis about DNA. Some who heard Lederberg's presentation described his results as "uninteresting" and "suspect." A few scientists took Hershey's results seriously, but then found ways to incorporate his findings into the protein paradigm. If Stanley were to restrain his somewhat prideful personality and show any interest in new scientific theories about DNA, it remained a scant possibility to the colleagues that knew him.[5]

Instead, he maintained his professional authority at the BVL and over the protein paradigm—and fully intended to use it. Swarms of bright young men, inspired by Stanley's reign over the most celebrated biological science program, his recent scientific successes, and his publicly displayed honors and awards, applied for much sought-after appointments. From these, Stanley selected staff as if he were preparing for war: heavily fortified with unlimited state funds, he promised key recruits the most powerful research "weapons" in the world, and organized everyone according to a hierarchy of rank. For instance, Stanley offered the chair of the virology program to his ex-lab partner and good friend Arthur Knight, whose personal and professional relationship developed while the two men worked at the Rockefeller Institute in Princeton during the 1930s. He bestowed a senior virology appointment on Howard Schachman, a chemical engineer, and offered him full authority to run one of the world's fourteen existing ultracentrifuges. Alongside his "virologists" he placed Gunther Stent, another colleague of Stanley's at Rockefeller and an emerging star in physical chemistry, who had achieved recognition at an early age as one of the leading figures in bacterial viruses. Stanley filled a large number of research assistant posts with highly skilled physical scientists whom he recruited away from tenure-track positions at other elite universities: Fred Carpenter, a protein chemist, left the University of Washington; nucleic acid chemist Charles Dekker, enzymologist Arthur Pardee, and carbohydrate chemist Donald MacDonald left the Rockefeller Institute at Princeton. A testament to the attractiveness of the BVL, Stanley filled positions that he considered "less-critical gaps" with prestigious short-term assignments: for instance, Rosalind Franklin, one of the leading crystallographers in the world, accepted two short-term research fellowships to work at the BVL. Stanley also kept a handful of immunochemists on a standby adjunct track to fill technical positions on an "as-needed" rotational basis.[6]

The newly recruited staff assembled by Stanley attracted almost as much attention at Berkeley as the physicists and their multiple laureates. The student newspaper cast these biological scientists as newcomers to the Berkeley campus, known and admired ivory tower men, whose enormous scientific potential garnered constant praise; their backgrounds and characteristics also made them objects of the scientific community's fascination. "Stanley's appointments," reported one leading science journal, "display a proper commitment to fundamental research." One of the BVL's reviewers determined that his careful selection of staff was "developing exceedingly well . . . and will most likely make great advances." Others anticipated that Stanley's appointments "would make the University of California pre-eminent not only in physics but in biochemistry as well." A local paper, the *Oakland Post Enquirer*, announced "Dr. Stanley's . . . laboratory will be as world renowned as is the radiation laboratory of Dr. E. O. Lawrence." Even though Warren Weaver at the Rockefeller Foundation categorically opposed Stanley's funding requests, he admitted in his diary: "There is no doubt that this is a splendid laboratory, in every physical sense, and an extraordinary group of people in it. One of the things that puzzle(s me) the most is that the interest to work with him in the laboratory seems extraordinarily high and they all speak in the warmest possible terms of Stanley. Since (I) consider Stanley a somewhat curious and complicated character, it is interesting and fine to see that he is able to command such enthusiasm on the part of all his staff."[7]

Amid these glowing remarks, a handful of longer-tenured biochemists began to note the constellation of star scientists that Stanley had assembled. They noted that his entire staff shared similar beliefs, in addition to their personal attachment to Stanley, four of which they found particularly significant. First, each embraced Stanley's commitment to virology and protein chemistry research. Second, many worked at some point in their careers alongside Stanley at the Rockefeller Institute in Princeton and had little, if any connection, to Berkeley upon their arrival. Third, a vast majority had extensive training in the physical sciences. Fourth, and perhaps most important, their collective interests implied a fierce commitment to fundamental research and a subtle hostility to what their staff identified as "applied concerns."

Committed to "liberating [the biological sciences] from its role as a service discipline," Stanley and his appointments made certain that research conducted at the BVL was "not beholden to a medical or agricultural school." Put another way, Stanley and his staff believed that applied research simply lacked theoretical vigor. Gunther Stent, one of Stanley's first appointments to the BVL staff, often mocked physicians who had ventured into their exclusive experimental labs and described their experiments as "contaminat[ed] by their M.D. spirit." Others would refer to visiting UCSF clinical researchers as "hat rack boys"—no one knew if they intended this as a derogatory phrase, but it certainly identified them as temporary visitors to an exclusive fraternity. Therefore, in Stanley's mind, and despite his promises to the California Legislature, experimentalists coming from an agricultural background—geneticists, agricultural biochemists, plant pathologists, and so on—or those who had clinical training—bacteriologists, immunologists, endocrinologists, and so on—need not apply.[8]

The decision to elevate fundamental—or pure—research left little room for alternative approaches in the BVL's churning scientific cauldron. Whether Stanley could contain and channel potentially disruptive practical interests, or whether his vision of a free-standing biochemistry laboratory dedicated to pure virology research would bow to them, nagged at many in the BVL. In this unstable, bipolar world, two staffing decisions in particular suggest the lengths to which Stanley would go to liberate his laboratory of the corrupting influence of applied research.

The first was a quiet and unassuming immigrant from communist China, Choh Hao Li. In 1935, Li left his chemistry program at the University of Nanjing and took a low-level laboratory assistant position in the Institute of Experimental Biology at Berkeley, in spite of the fact that his application to do advanced graduate work at the University of California had been turned down. One of many rank-and-file technicians, Li set himself apart by graciously completing the tedious task of isolating, purifying, and identifying the correct amino-acid sequences of anterior pituitary hormones that other technicians did everything they could to avoid. Most looked at the assignment as akin to finding and then counting over and over, day after day, beads on a string. Though endocrinology would in the next generation become a leading research discipline at all three Bay Area universities, most experimentalists did not use it as a base from which to launch a scientific career, because at the time, endocrinology was a field not clearly marked as either science- or medicine-based. On the one hand, the study of hormones is undeniably clinically relevant—all endocrine diseases, simply put, are the result of too much hormone or too little; on the other hand, hormone research requires a strong biochemical background, which warrants consideration as a hard and fast pure research topic. An ambiguous scientific commitment to fundamental research questions that might have practical relevance was a death-knell at a laboratory like the BVL that demanded total commitment to pure research.[9]

An exceptionally determined individual, Li soon mastered a field that others considered a mere stepping-stone that preceded more glamorous assignments. By the time Stanley took over the BVL in fall 1948, Li was the only scientist to have ever chemically identified a pure hormone, and he had done it four different times. Yet, despite his remarkable achievements, few would embrace a scientist whose field was neither pure research nor explicitly clinical. One investigator described the schism as "strongly divided between medical and non-medical [staff]" who rarely "talk[ed] to each other at all." Stanley's practice of prioritizing pure research over applied delighted nonmedical staff and administration committed to strengthening fundamental research in Berkeley's biological sciences, and ratified beyond their greatest expectations the wisdom of keeping separate from medicine. Basic researchers such as Howard Schachman, one of the BVL's ultracentrifuge artists, purportedly criticized Li for his "connections to the clinical side." On the advice of Wendell Stanley, President Sproul "severed [Li's] connections" to the BVL by creating and then assigning Li to the Hormone Research Laboratory, whereupon Chancellor Clark Kerr tried to "resettle Professor C H Li and his so-called 'Hormone Laboratory' as a demotion to the San Francisco Campus." With "no logical place to put him," Li soon found himself, his work, and his lab a forgotten entity in the growing Berkeley empire. Even notoriously generous federal agencies had little interest in Li's work, which forced him to rely on the financial assistance of less-than-orthodox private philanthropists, such as Mary Lasker, who the press at the time cast as a "monomaniac" Progressive reformer obsessed with practical matters like illness and healthcare. Left alone—an existence that this normally reticent investigator may have actually preferred—Li turned the small, isolated Hormone Research Laboratory into one of the most advanced endocrinology laboratories in the world, and received very little recognition for it.[10]

While the BVL staff and Berkeley administration pushed out Li's clinical interests, another biological scientist at Berkeley was curiously thrust into the netherworld that rested somewhere between pure and practical research. Horace "Nook" Barker, a plant pathologist who studied bacterial metabolisms under C. V. van Niel at Stanford, entered Berkeley's Agriculture Experiment Station in 1936 as an instructor of soil microbiology, but quickly gained notice in 1945 for synthesizing sucrose—ordinary table sugar—and then later for his ingenious use of radioactive isotopes borrowed from Lawrence's Radiation Laboratory to discover the active form of vitamin B-12. By 1946, Barker had such an impressive professional reputation that the Berkeley administration ironically took into consideration his support of Wendell Stanley to lead the BVL. Doubly ironic, Barker advocated the separation of pure and applied research at Berkeley. Little did Berkeley administrators or Barker know that his work with vitamin B-12 and his tenure in the College of Agriculture clashed mightily with Stanley's desire to create a pure research laboratory "not beholden to medical or agricultural interests." Put another way, Barker simply did not belong in Stanley's world of pure research.[11]

Despite Barker's many professional qualifications, Stanley demonstrated his intentions when he dealt a stinging rebuke to Barker's application to join the BVL. Scientific differences erupted between Barker and Stanley and forced staff to choose scientific alliances: Stanley advised subordinates to focus on virology, while Barker asked them to study bacteriology; Stanley emphasized molecular structure, Barker metabolic processes; Stanley wanted biochemists to sever all ties to agricultural topics, and Barker encouraged students working at the BVL to conduct comparative studies with the College of Agriculture. Imagined professional differences between pure and applied research had become reality.[12]

A quiver of foreboding crept into the BVL's vaunted status in the fall of 1951 when Barker sent a barrage of complaints to Berkeley administration about the overrepresentation of "Stanley's men" and the dearth of "biochemists from the Agriculture Experimental Station." But then Barker made an enormous tactical error. Instead of siding with Stanley's enemies at the hospital in San Francisco and preclinical investigators at

Berkeley, Barker tried to exploit the perceived difference between pure and applied research to his own advantage. In his letters to Berkeley administration, he cast his own work in agricultural research as just as pure as Stanley's, and therefore just as worthy of full access to the BVL. Much like Stanley, Barker did not believe that medical research deserved equal billing. According to Barker, the work of preclinical and medical staff was "quite unrelated to the rigorous demands of fundamental research," and Stanley's exclusion of medical researchers from the BVL as "understandable" rather than a missed opportunity.[13]

Armed with a few allies, Barker lashed at Stanley's favoritism, erupting into a cacophony of protest one year later. Staff from the College of Agriculture who had been barred from the BVL grumbled that Stanley's continued "emphasis on the physical-chemical aspects of biochemistry" left little opportunity for anyone outside his scientific inner circle. Plant biochemists such as Barker found especially distressing Stanley's obvious lack of interest in their work, and asked for an alternative program director, one who "knew something about (plant biochemistry)." Preclinical biochemists joined the fray, and argued that Stanley's decisions left them with little choice but to supplement their income with clinical consulting work, which added a deeper sense of disparity that staff who conducted medical research were "second-rate citizens." So obvious was Stanley's preference for "certain young men," commented one observer, that his choice of appointments left "the whole ineffective Life Sciences building in the somewhat peculiar position of being bypassed by much of the truly modern biochemistry and biophysics research . . . carried out at Berkeley." As much as Stanley tried to dismiss the complaints, his opponents constituted a real threat, since more biological scientists at Berkeley found the BVL's doors closed than found them open.[14]

Stanley refused to budge, so Barker stepped up his protest. He complained to President Sproul in a series of confidential letters—marked by a profound naïveté—that staff at the BVL had no teaching responsibilities. Not surprisingly, Barker's focus on teaching assignments backfired. Berkeley administrators considered research output more valuable than teaching and were in no mood to compromise their first real chance to supplant Harvard as the top research university in the country. Consequently, they threatened Barker that he might find himself with "less time in the laboratory," and then made an example out of one of Barker's allies, Leslie "Latty" Bennett, by assigning him five graduate and three undergraduate lectures per week, which left him with "virtually no time to work in the laboratory." A stunned Barker responded with an anemic defense: "as members of the Agricultural Experiment Station, our main activity has been research and I believe it is fair to say that we have been [productive]. In the past our teaching loads have been relatively light. If we are transferred to L & S (Letters and Sciences undergraduate curriculum) . . . our teaching load could be greatly increased at the expense of our research activities."[15]

Barker may have lost the battle against Stanley, but he was gradually winning over sympathizers who had grown tired of the administration's obvious favoritism. Biochemists left out of the BVL began to organize. Roger Stanier in bacteriology—another field expelled from the BVL—warned Barker that the war against Stanley for open access was about to become personal:

> Good luck in your dealings with Stanley. . . . I don't think he's actively and consciously malevolent in any of his dealings, but he is a stupid and muddled man in whose mind considerations of science, academic politics, scientific politics and personal prestige are all rather hastily and inextricably mixed. . . . He should never be left alone for too long, since if this happens the muddle takes over and before you know where you are, he will have embarked on an undeniable course from which he can only extricate himself by lying. . . . The worst difficulty is that he doesn't understand the climate of science and scientific discovery.[16]

Stanley and his allies in Berkeley administration took these polemics seriously and chose to ignore the more powerful scientific revolution now underway: the rise of DNA as the central paradigm in the biological sciences. Rather, they determined that the first and desperately urgent item of business was ending the political crisis waged by clinical and agricultural researchers. President Sproul, highly protective of his prized venture in biology, again pounced on the BVL's in-house critics, identified the disloyal, and then transferred them to less disruptive quarters. Indeed, Sproul reassigned so many qualified staff that he effectively created a series of overlapping research programs in the biological sciences, each marked by transience, instability, and a high rate of turnover: S. A. Peoples in the College of Agriculture was hastily relocated along with two other scientists to the University of California's Davis campus, a "farmer's" outpost eighty miles to the north in the San Joaquin Valley; long-time professors of biochemistry Herbert Evans and Edward Sundstroem were assigned emeritus status because both were about to "reach the age of normal retirement;" and much like Leslie Bennett, many other faculty suddenly found themselves designated "half medicine and half research, and then [spent] most of [their] time commuting between [UCSF and Berkeley] on two different sides of the Bay." As for Barker and his plant-pathologist allies, Sproul sentenced them with the most vindictive punishment available at a research university such as Berkeley: he banished them to the bowels of the College of Letters and Sciences and assigned them to teach "insufferable" introductory courses to undergraduates. Sproul's house-cleaning left Stanley with a homogenous, united core of extremely talented investigators and his critics wondering if they could ever penetrate the forbidding walls of Stanley's BVL.[17]

Confident, safe, and buoyed by the protection given him by university administrators, Stanley could now direct all of his attention to reestablishing the supremacy of the protein paradigm. His preferred method also remained the same: clues to unravel the mysteries of life were hidden deep inside the virus. Therefore, Stanley organized large numbers of experimentalists into teams, equipped them with every available technology imaginable, and then focused their efforts on a single scientific question—then he established equally formidable "counterfactual" experimental teams to disprove the stated objectives of the first team. One example of a working experiment at the BVL looked like this: it was unclear at the time whether protein molecules, known to be made up of a very large assortment of a small number of components—the twenty-odd amino acids—were jumbled together like mixed nuts in a bag or were orderly, and whether their physical structure had a functional purpose. To answer this fundamental bioscience question, Stanley would organize one team of scientists as a "covalent bond group," which posited that proteins might serve as a structurally sound skeletal mass—much like the honeycomb of a beehive; in contrast, a counterfactual team of "numeric theorists" sought mathematical patterns in a protein's structure that might identify a functional or metabolic purpose within the jumbled mass. Stanley also encouraged investigators at the BVL to use Lawrence's Radiation Laboratory to strengthen their physical and chemical studies of the tobacco-mosaic virus. Russell Steere frequently collaborated with physicists and chemists at the cyclotrons to conduct research on the chemical composition and physical structure of plant viruses. Dean Fraser collaborated with cyclotron chemists to complete a series of procedures using radioisotopes to confirm the synthesis of bacterial viruses. And significant numbers of Stanley's staff used Lawrence's radioactive isotope labels to study metabolism.[18]

Unquestionably, Stanley's leadership and scientific focus dramatically increased the experimental productivity of the biological sciences at Berkeley. With a world of talents and resources so vast and deep—physical, technological, financial, and intellectual—investigators at the BVL hammered away at the physical structure and chemical composition of proteins, and they left all other experimental topics, such as DNA, sputtering in their wake. Using the latest experimental techniques and most powerful technologies available, they reported on protein's astounding diversity and import, and they chose to ignore the repetition of evidence that showed the biological significance of DNA. Their extraordinary successes helped establish the disciplinary parameters for protein chemistry and protein-sequencing. This result was, in fact, Stanley's experimental and professional objective. From such a powerful research laboratory, what else would BVL bioscientists have to do to confirm that protein was the central force of all life? From such brilliant minds, what did BVL bioscientists have to fear?[19]

Stanley may have been a magnet of controversy at Berkeley and a target of frustration for those few who had dabbled in DNA research, but to everyone else remotely familiar with the biological sciences, the BVL was a roaring success. Student enrollments surged in programs that Stanley promoted, while enrollments fell in those programs that Stanley disparaged. The number of undergraduate students signing up for microbiology more than doubled. Biochemistry absorbed three times the normal enrollments at its height. Biophysics enrollments quadrupled. Those trends indiscriminately depleted medical and agricultural programs, rendering some, such as immunology and plant pathology, to struggle for their organizational survival. Private patrons like the Rockefeller Foundation reluctantly acknowledged that the BVL had become the "leading [program] in biochemistry in the country." The public, too, clamored for their attention, especially the ever-present Stanley; never one to shy from the limelight, he tried to accommodate as many speaking engagements as possible. More significantly, however, the future seemed secure as research patronage poured into the BVL coffers. Federal money came from, among other agencies, the United States Public Health Service and National Institutes of Health, the Atomic Energy Commission, the Office of Naval Research, and the United States Department of Agriculture. The California State Legislature continued to provide support for the program as well. Private donations also expanded, coming from organizations like the American Cancer Society, Merck, and Sunset Magazine, and from quite a few wealthy individuals too. All of this money went back into Stanley's lab, paid salaries of administrative and support staff, reimbursed general university funds, or provided graduate students and postdocs with more fellowship money.[20]

It was late 1952, and Stanley had successfully withstood numerous attempts to topple his grand experiment. The Berkeley administration felt, at long last, that they had gained a firm hold on the field. But no one seemed to note that the biological sciences were heating up. Or perhaps no one yet cared.

History usually turns slowly. Changing social consciousness or shifting economic conditions can convince people to accept an entirely new set of ideas, but this is the kind of history that happens slowly. Science policy grew slowly in the years after World War II. So did the relationship between scientists and the state.

Sometimes, however, history turns on a pivot.

On 25 April 1953, *Nature* published a modest communication from two researchers at the Cavendish Laboratory at the University of Cambridge, England.

> We wish to suggest a structure for the salt of deoxyribose nucleic acid (D.N.A.). This structure has novel features which are of considerable biological interest . . . it has not escaped our notice that the specific pairings we have postulated immediately suggests a possible copying mechanism for the genetic material.
>
> Sincerely,
> James Watson and Francis Crick

About one month later, Watson and Crick published another article in *Nature*, this time asserting with much more confidence some of the genetic implications for their proposed model of DNA:

> Any sequence of the pairs of the bases [A:T, C:G] can fit into the structure. It follows that in a long molecule many different permutations are possible, and it therefore seems likely that the precise sequence of the bases is the code which carries the genetical information. . . .
>
> Our model for deoxyribonucleic acid is, in effect, a pair of templates, each of which is complementary to the other. We imagine that prior to duplication the . . . bonds [connecting the bases] are broken, and the two chains unwind and separate. Each chain then acts as a template for the formation on it itself of a new companion chain, so that eventually we shall have two pairs of chains, where we only had one before. Moreover, the sequence of the pairs of bases will have been duplicated exactly.

The two articles stunned biological scientists. What was at stake, many instantly recognized, was the fate and future of their field. What is the proper reaction to a new scientific paradigm? How does one dispose of the protein paradigm overnight and embrace Watson and Crick's momentous proposition that DNA might be the "code" of life? More importantly, *should* investigators dispose of the protein paradigm? Certainly, nothing prevented a scientist from changing to a different experimental topic; virtually any strong research proposal—even for work on a seemingly inconsequential topic as DNA—had a high probability of receiving support from any number of federal agencies. That was, ultimately, the purpose of new science policy: to provide investigators with financial security so they could take scientific risk.

Yet at Berkeley, as elsewhere, the discovery of the structure of DNA created a local crisis that required an astute response. Until Watson and Crick proposed the double-helical structure and internal copying mechanism for DNA, Stanley's only threat came from wretched conflict between academic colleagues. Could this recent scientific discovery, many wondered at the BVL, though surely welcome for its own sake, destabilize their professional authority? Stanley's illustrious career, as well as the professional and personal fate of many within the BVL, hinged on how their leader would react: should they follow the most recent shift in scientific theory, or should they continue their triumphant work in traditional projects that they had only recently begun?[21]

The Ordeal of the BVL

Watson and Crick's articles in *Nature* pushed Stanley to the precipice of an epic personal and scientific conundrum. He had plunged his staff into the promising field of protein research, only to find it yanked out from under his feet. In a letter to a colleague overwhelmed by the hysteria over DNA Stanley fatefully cautioned: "I am rather inclined to believe that the importance of . . . the (DNA) findings is somewhat overemphasized." A few curious scholars in the BVL looked anyway. Robley Williams took a few electron microscope photographs and x-ray diffraction pictures of molecules of the tobacco-mosaic virus' ribonucleic acid (RNA) and wondered about their dimensions and density. Heinz Fraenkel-Conrat began a painstaking experiment of taking apart the whole tobacco-mosaic virus and reassembling it, but this time with an eye for a "DNA point of view." Both men wondered about the magnitude of the laboratory, the growth of the staff, and the centrality of protein research—an awe-inspiring combination that had created a self-perpetuating and self-fulfilling system. Fraenkel-Conrat worried that he and his colleagues in the BVL had been, for the most part, too slow to catch on to DNA.[22]

Desperate to stabilize his laboratory, disappointed at the timidity of the community that had until recently considered protein research the paramount experimental topic, and faced with his own unwavering demand for "scientific excellence," Stanley made the courageous decision to shift course and have staff conduct research on the physical and chemical properties of proteins *and* DNA. His scientific somersault came too late. Barker and his rebellious biochemistry allies had already established a plan of attack on Stanley's vice-grip on the BVL and his decision to exclude biochemists with connections to preclinical or agricultural research. Stanley thought himself secure, at least, that earlier rebellions had been put down, an autonomous biochemistry department had been organized, an independent Virus Lab had been established, the public overwhelmingly liked him and supported his work, and university administrators had stood behind his every decision. And much science remained undiscovered. Discovery was, in the end, everyone's primary commitment.[23]

It soon became painfully obvious, however, that Stanley's original decision to emphasize protein research had enormous and lasting consequences. Once-generous private foundations and corporations no longer found the BVL a worthy recipient of their donations. Lederle Labs, which once clamored to give large sums of money to the BVL, "regretfully decided not to renew the . . . grant in support of . . . work done at the BVL." Merck, which a year earlier had given $25,000 in soft-money support, wrote during the BVL fallout following Watson and Crick's discovery that "the general . . . situation . . . has led to an extensive revision of our scientific program . . . it is, I am sure, unnecessary for me to elaborate any further, so let me just say that we are going to have some difficulty in continuing support for your . . . research program."[24]

The decision by private sponsors to recall or terminate their support stunned Stanley. Professional humiliation soon followed. George Gamow, an iconoclastic astrophysicist and a colleague of Stanley's at Berkeley, challenged the entire scientific community to find "how the four bases of DNA—adenine, thymine, guanine, and cytosine—could be arranged to specify the assembly of 20 amino acids which then construct an infinite number of proteins." To lead the search, Gamow established an exclusive RNA Tie Club, made up of twenty of the world's leading biological scientists—one for each amino acid—and he left Stanley's name off that list.[25]

As Stanley and his colleagues tried to slow the unraveling of the BVL, Horace Barker and his biochemistry allies staged their most aggressive campaign yet to oust Stanley from his departmental chairmanship and authority in the Virus Laboratory. Some complained that Stanley had compromised the BVL's scientific objectivity by accepting donations from private companies, a claim that seemed sadly hypocritical considering that Stanley also accused plant pathologists in the College of Agriculture and preclinical researchers in the UC medical school of harboring "applied concerns" that compromised their own work. But this time Stanley was on the defensive. To soften the blows delivered by his critics, he graciously offered to give them access to the third floor of the BVL, a deal they rejected and then responded with an even more strident demand that Sproul replace Stanley at once.[26]

Then came a series of personal attacks, highlighted by one in particular, delivered at a holiday party, in which staff members staged a vindictive skit, an epithet that Stanley did not appreciate, which they sung to a tune from *HMS Pinafore:*

When I was a lad served three
As a graduate student on my Ph.D.
I wore a white coat and a winning smile
But I didn't know anything all the while.
I hid what I didn't know so gol' darn well

Of scientific jargon I acquired such a grip
That while talking shop rarely made a slip.
I expounded all day on things I didn't know
Knowing all the while that it would not show.
This made my lectures clear as a bell
So now I am director of the BVL.

I became so great that it occurred to ME
That I really might have ability.
I thought for a while that I should have spent
At least one day on an experiment.
I never did but its just as well
For now I am director of the BVL.

Now students all whoever you may be
If you want to gain success like ME
Be sure and never go near a lab
For all you really need is the gift of gab.
I have this gift as you can tell
For I am the director of the BVL.[27]

At this critical juncture, amid growing hostility, Sproul deserted Stanley, whom he had guided so surely for his first five years at Berkeley. Battered and beaten by an intense anticommunist loyalty-oath controversy that he had also profoundly mishandled, Sproul distanced himself from the melee in the BVL by notifying combatants that they would have to find an "amicable solution on their own." Behind closed doors, however, Sproul tried desperately to right the sinking ship by filling research holes left vacant by Stanley. He even asked Horace Barker for suggestions. News of the BVL's turmoil spread so fast, however, that two of Sproul's top choices—Arthur Kornberg, a biochemist at Washington University Hospital in St. Louis, and Joshua Lederberg, a geneticist at the University of Wisconsin at Madison—rejected informal overtures because they had heard "things with Stanley were going no where, if not getting worse."[28]

Making Sense of the BVL Debacle

Berkeley would retain for years many of its highly skilled investigators, maintain the autonomy of the new biochemistry department and the Virus Laboratory, and receive generous amounts of money from federal agencies to support all of their programs in the biological sciences. Nevertheless, the curtain had fallen hard on the BVL. While scarcely profound or unreasonable, the dreams and expectations for the BVL created illusions that ran deep. Cultivated by the press, nourished by determined university administrators, and fiercely protected by its own investigators, the BVL left observers then and scholars now alternately confused and perplexed, caught up in the puzzle about why such a promising facility as the BVL would have such a difficult time becoming a leading research facility.

Many analysts, including some who worked at Berkeley during the 1940s and 1950s and who helped shape opinions in recent years, have always been convinced that the decisive factor in the fall of the BVL was Wendell Stanley himself. They conclude that responsibility for the BVL's limitations—in particular, the decision to ignore DNA—rests squarely on the shoulders of Stanley and his obsession with pure research. And they point to Stanley's egotistical personality as the reason for his fatal reluctance to depart in any significant measure from the obsolete dogma of the protein paradigm. It is no exaggeration to say that most critics—then and today—agree with the opinion of one scholar that Stanley's decision to dismiss DNA was a "depressing . . . failure."[29]

As much as Stanley mismanaged the facility, it is possible that these critics have overstated him as the villain of the BVL. No one forced researchers at the BVL to direct their experimental topics to coincide with Stanley's scientific interests. Rather, investigators there generally found professional appeal and scientific justification in Stanley's unrelenting commitment to protein research, and they appreciated having access to the best equipment in the world, seemingly limitless financial support, and fast-paced, outcome-oriented investigation focused on a single experimental question. The work also provided professional opportunity for ambitious newcomers to the field looking for a subdiscipline in which they could develop their own expertise. Together, their efforts lent a sense of being part of a large and powerful team, and professionally, their work promised to reestablish protein as the most important analytical framework in a field that many believed had become overly enamored with unsubstantiated claims about DNA and its central importance in life.

To these considerations it must also be added that perhaps UC administrators, such as President Sproul, and even the California State Legislature for that matter, should not have given Stanley the means to run a highly autocratic, interdisciplinary research program at one of the country's preeminent and democratically predisposed public university. Berkeley-style politics and the propensity of the staff to embrace debate tore at the cohesiveness of the BVL and undermined its integrity; perhaps young research laboratories like the BVL in ever-shifting scientific fields like the biological sciences were especially vulnerable to the wicked play of UC politics. And it must also be repeated that all of these individuals faced difficult choices at about the time that Watson and Crick made perhaps the most significant discovery of the twentieth century.

One of the great challenges in writing about scientific discovery in historical perspective is the temptation to judge past scientific experiments from the present state of knowledge. "Presentism," more commonly known as hindsight, corrupts the uncertainty that defines the process of scientific research. The staff at the BVL was neither alone in its dedication to the protein paradigm, nor was it the slowest to recognize the centrality of DNA. Many academic research programs overcommitted to the protein paradigm in the late 1940s and early 1950s had difficulty responding to mounting evidence that DNA was the principal component of each cell of every living thing. These institutes include CalTech, the Rockefeller Institute, Vanderbilt, Oak Ridge, Cold Spring Harbor, King's College Cambridge and London, or the Pasteur Institute.[30]

Angela Creager, a historian of science, analyzes the difficult decisions faced by biological scientists at the BVL. Creager courageously downplays Stanley as a disruptive personality. Instead, she uses the experiences of Stanley and the BVL as a case study to highlight the "disciplinary ruptures" caused by the "demarcation between biochemistry and molecular biology." As Creager points out, it would be unrealistic to expect any scientist to have a "prescient sense" of DNA's remarkable future, and no amount of money could buy that sense either. If Watson and Crick's discovery of the structure of DNA really was revolutionary—and by all accounts, it deserves this vaunted status—then Creager offers a compelling defense that Stanley faced a near-impossible task of redirecting a young and underdeveloped research laboratory during a scientific revolution that had reconfigured traditional disciplinary boundaries.[31]

Creager's approach offers a useful way to understand the biological sciences during the late 1940s and early 1950s. At the same time, however, Creager's argument creates fresh difficulties. Can we say, for instance, that the BVL struggled because it was caught in the chaos of a developing discipline? Indeed, other laboratories eventually pulled through and even made important contributions to the field. Does a broader view merely return us again to an internal story of tragedy and personal failings? Or was it something else, in addition to Stanley's failings and the wrenching confusion of a new biology, that might have caused the disciplinary rupture at the BVL?

A clearer view of the challenges emerges when the historical uniqueness of Berkeley's biological sciences is considered in the context of the scientific revolution brought on by Watson and Crick's discovery of DNA's double helix and internal copying mechanism. From this vantage point, what stands out is the tendency of bioscientists at the BVL to embrace inflexible local management styles and rigid opinions about what constituted proper research. Furthermore, these stubborn commitments intensified at a time when new scientific ideas required more flexibility, not less. No doubt the BVL struggled because investigators did not fully appreciate the "disciplinary rupture" caused by Watson and Crick's discovery, as Creager points out. But this aspect is but part of the full story.[32]

The case of Wendell Stanley and his management of the BVL merely shows in exaggerated form an unfortunate tendency of academic programs to cling at critical moments to rigid disciplinary or institutional boundaries that inadvertently constrain new ways of thinking and decenter potentially dynamic projects. The divide at the BVL—between Stanley's virology, Barker's plant biochemistry, and Li's clinical endocrinology—actually reflected the general tendency to separate pure research from applied, an irreparable social climate that slowed research efforts at the BVL. Stanley's obvious social incompetence notwithstanding, open interdisciplinary research stopped at the border of collaboration when a potential participant was deemed to harbor practical interests that corrupted the perceived objectivity of scientists conducting pure research.

It was not disciplinary rupture that thwarted an otherwise worthy effort of the BVL, therefore, but disciplinary rigidity. In general, rather than confront and adapt to new scientific discoveries, investigators at the BVL tried to direct them, believing in a sense that they had to control the intellectual foundations that supported their work. To anyone remotely familiar with the built-in dynamism of science and the unpredictability of the experimental process, the attempt to define, direct, or control scientific discovery, though certainly understandable, is a profound impossibility.

Watson and Crick's discovery required disciplinary and institutional flexibility. But UC administrators and the BVL's management and staff tried to hoist stability and security upon the protein paradigm and ignore DNA, which ultimately—and ironically—caused disciplinary ruptures that they so desperately wanted to avoid. In this particular instance and at this particular point in time, the BVL's rigid disciplinary habits prevented the program from capitalizing on its first-mover advantages, while underdeveloped programs at Stanford and UCSF could approach the biological sciences from a clean slate.

The fate of the BVL offers one more lesson: the boundary that separated pure and applied research could inspire bioscientists of this era to achieve heights once thought unapproachable; who then could have possibly imagined that the weakening of this same boundary might inspire the next generation toward another, still unfinished, scientific revolution?

Chapter 4

The Ascent of Pure Research

A medical school's goal should not be to turn out general practitioners. Instead we expect to produce well-trained basic medical men able to go on to further breakthroughs.

—Robert Alway

I suppose the M.D. [degree] ought not to be held as prima facie *evidence against the possible qualifications of a postdoctoral research fellow.*

—Joshua Lederberg

The experience of Berkeley and the BVL only partly explains the rise of pure research in Bay Area bioscience laboratories. To review, pure research at the BVL forged ahead because the federal government allowed Stanley to establish standards of eligibility and payment. Stanley possessed, in effect, the power to execute the will of his own research agenda. Except the experience of the BVL also shows that generous federal policy did not necessarily guarantee success; indeed, federal money could antagonize preexisting professional tensions and sustain, for a time, fragile research agendas.

Nevertheless, biological scientists continued to cultivate pure research assiduously—a focus of this chapter. Stanford's bioscientists faced an organizationally complex dilemma: the university had a variety of unrelated research programs scattered throughout the life sciences. The challenge at UCSF was also quite different; here was a medical center run by doctors who allocated resources according to the needs of their patients, while scientists interested in conducting pure research had to show how their experiments might contribute. Berkeley bioscientists championed a simple solution for the BVL debacle: establish a program more intensely devoted to pure research than had been attempted theretofore.

Straining out the differences, organizational and institutional, the narrowing of the biological sciences at Stanford, UCSF, and Berkeley—and by extension, the ascent of pure research—occurred because of the earnest and at times obsessive determination of university administrators and faculty at all three universities. The channeling restraint of applied research was removed from larger strategies because compromise and accommodation were rarely given a high priority. In other words, pure research took hold in the biological sciences because bioscientists were willing to hoist their agenda on to or around those who opposed them. The unilateral efforts to push pure research also foreshadow the vulnerability of unresponsive science and its fleeting tenure.[1]

Unlikely Beginnings

Despite the exhilaration of the Berkeley community for the biological sciences, despite the exertions of Wendell Stanley, despite the efforts of the faculty, despite all the ingenuity and exuberance of Stanley and his colleagues, the BVL did not take. The tonic effect of Stanley's early declaration that "the BVL was destined to become the top bioscience program in the country" had long since worn off. To many of those who had put their faith in Stanley in 1948, and especially to those in Berkeley administration who had hoped for something more dramatic than prudent discovery and piecemeal publications, the BVL appeared, even before it reached its fifth year, to be a spent scientific laboratory.

Instead, the forward march of the biological sciences in the Bay Area began at Stanford University, though no one could possibly have imagined it at the time. The university had been launched by Leland Stanford, founder and president of the Central Pacific Railroad and former governor of California, and managed by his wife, Jane Lathrop Stanford, in memory of their son, who had died at age fifteen. The Stanfords considered the university as their own personal operation, and refused to accept any money from outside sources or donors. Typically encumbered by debt and legal difficulties, the Stanfords limited the university to a few hundred faculty and several thousand students, all of which was surrounded by—dwarfed some would say—a huge expanse of undeveloped land.

Stanford was still a spacious and unremarkable university when Frederick Terman arrived as the new provost in 1946. With an eye on Berkeley, whose greatness had already been established by the end of World War II, Terman wanted Stanford to become an elite university too. This vision meant that faculty should do a lot more research, and he turned to physics, the computer sciences, and engineering as targets for reorganization, where his policies contributed to the creation of Silicon Valley.

Biological scientists proved far less accommodating of his overarching strategies.[2]

No one in science should stand by idly, Terman reasoned, so he began relaying modest signals of encouragement and opportunity to the biology faculty that he had inherited. He asked them to teach fewer courses, and he offered to upgrade their "underprivileged" experimental equipment and hire more research assistants to help them with their experiments. But when Berkeley opened its new biochemistry department and Virus Laboratory, and then hired Nobel laureate Wendell Stanley to run them, and Stanford's biologists continued to neglect emerging experimental directions, Terman could no longer restrain his ambitions. He lashed out, sulfurously condemning the entire program. "A miserable disappointment," he called it. "They offer little value to the university," and he demanded immediate reform on pain of "eventually be[ing] disbanded." The remarks were cryptic, but they sum up, in a hostile way, Terman's anxious administrative style.[3]

Stanford's tweedy group of biologists could not understand Terman's disappointments, and all looked upon his attacks as meddling and unfair. Their go-it-alone approach to research had generated results, and no one could question their commitment to student instruction. And of course, they knew that few research universities could match the three celebrated life-science programs already in place at Stanford: a biochemistry research unit run by chemists on the main campus in Palo Alto; a medical center an hours' drive north in San Francisco; and a classical evolutionary and naturalist marine biology program at the Hopkins Marine Station in Monterey Bay, approximately one hundred miles south of Palo Alto. The collective temper of these three programs was magnified in its leading experimentalists: Hubert Loring, George Beadle, and C. B. van Niel.

Hubert Loring, one of four chemists practicing biochemistry on the main campus, arrived at Stanford in 1939 with a reputation as a private and conscientious experimentalist firmly committed to the idea that academic research was a profoundly individualistic endeavor. He disliked conformity—some said competition and scrutiny—and did not appreciate fashionable branches in science. In describing his own research style, Loring admitted, "if anybody began to work in something I was working at I would drop it and turn to something else." Withdrawn from the larger network of investigators, Loring developed conspicuously unorthodox opinions about nucleic acid that actually preceded the direction of biology by about five years. "Contrary to . . . the older idea of nucleic acid as relatively simple [and of] unknown function," Loring suspected that "the nucleic acid molecule [may have] sufficient complexity in chemical structure to account for chromosome specificity."[4]

George Beadle, Stanford's leading geneticist, had left CalTech years earlier in part because he felt that the "applied research" that his colleagues did for farmers in the San Joaquin Valley "was too slow," which to him had tainted the purity of academic science. Naturally, the isolation of Stanford appealed to him because there was less pressure to do research that had relevant meaning. While at Stanford, Beadle recruited E. L. Tatum, a prominent biochemist, and together the two formed a highly productive partnership. Indeed, in 1942—at the height of the war—Beadle and Tatum perfected an experimental system that induced genetic mutations in *Neurospora*, a theoretical discovery that helped establish the one gene–one enzyme rule for molecular biology and would later earn them both a share of the Nobel Prize. Beyond the realm of pure research, pharmaceutical firms such as Merck, Squibb, Sharp and Dohme, and Lederle Labs took note of the new experimental system too. Used on penicillin, Beadle and Tatum had found a way to increase antibiotic production levels *and* effectiveness; the difference here was not scientific knowledge but an extravagant commercial opportunity. Fueled by competition, pharmaceutical companies began dueling for licensing rights to the new experimental system, but Beadle and Tatum balked at such unwanted attention, and they said so in private correspondence: "[We have] no interest in a patent or any personal profit." To Beadle and Tatum, the most significant element of their discovery was not money but something else: the needs of war. In a lightning demonstration of their ideals, Beadle and Tatum waved all licensing rights and gave the OSRD authority to coordinate production of penicillin through the Department of Agriculture. With obvious pride, Beadle and Tatum had confirmed their own image of heroic academic science: "We can now go ahead with our work with clear conscience." It is one thing to adopt a worldview of science predicated on the conviction that knowledge is a public good and not for private gain. It is quite another to actually live it. Later generations would struggle with this same dilemma.[5]

Stanford's unquestioned leader in biology was microbiologist C. B. van Niel, whose many years off campus at the Hopkins Marine Station bred in him an acute appreciation for privacy. Isolated and alone, van Niel studied without interruption nonpathogenic bacteria—"little beasties" he called them. His colleagues referred to him as "King of the Microbes," a designation that seemed to one admirer as "too ordinary for such a god-like creature." That praise may have been exaggerated, but van Niel's professional accomplishments were by any measure extraordinary: by the time that Terman arrived in 1946, van Niel had already shown that some marine bacteria utilized light and carbon dioxide to produce their own food, published numerous monographs on bacteria, and gathered one of the largest private bacterial-culture collections in the world, numbering well into the thousands. Van Niel's thirty-year commitment to nonpathogenic bacteria also confirmed for his colleagues that Terman had underestimated the productive possibilities of self-directed research.[6]

But considering Terman's administrative objectives, his frustrations with Loring, Beadle, and van Niel were reasonable and unprovoked. An electrical engineer by training and son of Lewis Terman, the noted developer of intelligence tests, the younger Terman brought to Stanford a Progressive's belief in rational organization and an enthusiasm for social engineering. He was fascinated with science, firmly believed that it was the supreme academic field, and was openly contemptuous toward anyone who disagreed. It was in Terman's character to admire active and productive scientists, and he was driven to make the biological sciences more prolific. While Terman's ideas were clearly influenced by his father, he would later champion federal science policy or, more precisely, federal patronage that supported scientific research, as the impetus that launched his impassioned drive—and the unmistakable remedy to invigorate the biological sciences at Stanford.

Terman may not have understood the existing staff's preference for professional isolation, but he astutely recognized three critical features that prevented him from implementing his vision. First, the geographic isolation of all three programs prevented collaboration. Second, the distance that separated these programs fostered an ideology and a culture that had become deeply implanted within the faculty. Terman considered these two shortcomings a single problem: parochialism—whether as disdain toward softer clinical sciences, as fear that medicine's commitment to patient care might squeeze laboratory research, or simply as faith in the established boundaries that separated biology into discrete disciplines. Unarguably, Terman determined to force the biological sciences out of their fragmented and insular communities by reorganizing departmental relationships, but neither Loring, Beadle, nor van Niel showed any willingness to cross intellectual or institutional boundaries.

More frustrating to Terman, however, was the faculty's parochial aversion to federal patronage for scientific research. On the one hand, the highest ideal of Stanford's tradition-bound biologists was the protection of self-directed research, thus privileging professional autonomy. On the other, Terman's overriding goal was research output, thus privileging experimental productivity. To promote the latter, Terman spoke of tapping the government to pay for large research teams and expensive experimental equipment, an unimaginable proposition for faculty who believed that money to support research created unavoidable conflict of interest that threatened their highest professional ideal: scientific objectivity. Indeed, Terman may have considered federal patronage a "win-win-win" relationship for the university, its faculty, and the public, but Stanford's biologists stubbornly clung to their professional autonomy. It remained to be seen whether their independence would prevent them from uniting against Terman to defend their own scientific approach or from contributing to the DNA revolution just under way.[7]

A meeting of Stanford's Board of Trustees in late May 1953—just two months after Watson and Crick's momentous discovery—provided Terman the occasion to begin his quest to reorganize the biological sciences. Stanford's president Wally Sterling spoke fully and specifically for Terman: "We have a Medical School problem," said Sterling with great urgency. "Medical education, which is now in a state of flux," noted Sterling, "is inextricably tied to the basic sciences." "The key is the relationship of medical education . . . to other scientific fields." Where, exactly, was Sterling leading, more than a few trustees wondered. The announcement—that the San Francisco location of the hospital was inconvenient, that faculties were slow to collaborate, that student enrollments were declining, and that other medical schools had integrated a decidedly scientific component into their curriculums—may have been discomforting, but it was certainly something they had known.[8]

Terman had in fact been planning Sterling's message for more than a year, in letters to resolute faculty, in public presentations to alumni, and in articles published in the student newspaper. Despite the frequently repeated accusation that Terman lacked a coherent plan and had no capacity to reorganize the biological sciences, Sterling's presentation to the board had etched at least the ideological heart of a larger objective. Then Sterling delivered a stunning solution: "bring . . . the Medical School into the closest possible physical and intellectual relationship to the whole University."[9]

Neither medical training nor patient care was the driving issue behind Wallace Sterling's proposal to move the hospital from San Francisco to the main campus in Palo Alto, and Stanford faculty knew it. So did local communities. Physicians at the Stanford hospital in San Francisco condemned the proposal in a petition and vowed a fight to protect their jobs in the city. The Palo Alto City Council followed the lead of Shirley Temple-Black and threatened to veto all of Stanford's applications for building permits until the university gave some indication that the new medical school would provide health care for local communities. On campus, a significant number of students spoke out forcefully too, providing a glimpse of the friction to come in the next generation. "Why must part of my $600 tuition payment pay for a professor whose antiquated lectures do nothing but satisfy academic requirements?" asked one student rhetorically. Another reiterated a similar theme: "When a university overemphasizes the function of research, doesn't it risk losing sight of its original end—the preparation of successful and useful citizens?" When asked about the ideal purpose of biology, another pointedly remarked: "human betterment."[10]

Stanford's plan for a "Bio-Medical School," at first seemingly a sure thing, stalled before it could get off the ground. University administrators were stunned, uncertain what to do next. This rare opportunity to stretch the biological sciences, a plan that once offered so much promise, was now in full retreat. Even more dispiriting to proponents of the Bio-Medical School was the source of the opposition, a suspicious, provincial, ignorant public—in private, Terman and his allies derided critics as "cranks"—yearning for simple explanations of medical problems and immediate remedies for complex illnesses. Nevertheless, by late 1955, Stanford administrators had accepted the obvious: their attempt to promote the biological sciences in the medical school had split the community from the university, menaced the faculty, and endangered their larger goals.

The thunder rolling up from public quarters prompted university officials to begin considering in earnest dramatic revisions to the biomedical hospital, not for the ideal of scientific research education, but for the arithmetic of public support. That meant reversing their original priorities and fitting a medical center with the public's desire for clinical care, a tact they might have pursued from the outset. Dean of the medical school Windsor Cutting embraced student concerns by promising course requirements with "fewer formal scientific hurdles . . . than any other American medical college." To the great relief of staff in San Francisco, Stanford administrators also pledged to keep that hospital open and even promised to spend an additional $1 million to improve the facility. Finally, to assuage the concerns of their most formidable opponents in Palo Alto, the trustees proposed a new five-hundred-bed research hospital on campus. It was plain that no one at Stanford was at all impressed with the scientific rigor practiced in medicine, but even if university officials had actually wanted to operate *two* hospitals, they did not have the funds to do so. Thus, in a bizarre turn of priorities, Stanford signed an agreement with the city of Palo Alto to jointly build and operate a community hospital.[11]

On 8 August 1956, the university called on its most distinguished alumnus, former President Herbert Hoover, to announce their historic agreement with Palo Alto. Calling it "the largest effort of its kind ever undertaken in the West," Hoover compared the joint Stanford–Palo Alto Medical Hospital with his own work as the "Great Humanitarian" during World War I. In the end, though, more than a few commentators found the whole agreement uninspiring, including one who noted an eerie similarity between Hoover's experience in the White House during the Depression and Stanford's surreal and inherently unstable compromise, a patchwork of scientific and clinical concerns that pushed dangerously near insolvency. With the Palo Alto mayor by his side, Stanford University President Wallace Sterling proclaimed the agreement as a "great uniting of town and gown," but a mighty host of critics quietly assailed the joint hospital as "Wally's greatest folly." Either way, Stanford's difficulties in remaking the biological sciences occurred at a critical moment in the history of science. The Soviets were about to launch the world's first rocket ship.[12]

The Rebirth of Pure Research at Stanford

On 4 October 1957, while Stanford and Palo Alto struggled with the details of hospital design, history turned the biological sciences on another immediate pivot when the Soviet Union announced the successful launching of *Sputnik*, a basketball-sized satellite that traveled over the United States as it orbited the Earth. A month later the Soviets launched *Sputnik II*, a much larger satellite that carried a variety of scientific instruments and, adding further insult, a dog. The United States responded with its own artificial satellite—a Vanguard missile—but it rose two feet off the ground, crashed, and then burned.[13]

The Soviets, it seemed, had conquered space; Americans, by and large, panicked. Elected officials questioned the axiomatic assumption that its liberal institutions were "better" and its military "mightier" than their chief rival, and the public quivered in fear about the intrinsically more meaningful rocket that the Soviets had launched and the threat that it posed to their security. Significantly, *Sputnik* served as the goad that strengthened the relationship between government and science. No longer did policymakers see science as a spontaneous and separate entity. Rather, elected officials took it upon themselves the responsibility for generating scientific discovery. Eisenhower quickly mobilized the President's Science Advisory Committee and gave them direct access to the White House. Not to be outdone, Congress accused the president of not doing enough, and then appropriated massive increases of funding to various sponsoring agencies. Most budgets that supported science tripled overnight, such as the NSF's support of basic research, which increased from $50 million to $133 million, while appropriations for the NIH surged at a faster rate, from almost $25 million just before *Sputnik* to $135 million.[14]

Much indeed had changed in America, but much remained the same too. Both political parties hammered away at each other for losing the space war, agreed wholeheartedly in strengthening science policy, and left intact the mechanisms through which federal money was distributed—biologists still occupied the most influential positions within each agency. Moreover, *Sputnik* had done little to relax the opinion in the biological sciences that saw pure research as superior to and incompatible with applied. Rather than slow momentum, *Sputnik* had galvanized the field.

Intoxicated by the new mood of the country and the generous science policy that it engendered, Stanford administrators unilaterally severed their agreement with San Francisco to manage the university's hospital and their agreement with Palo Alto to build a 500-bed hospital. Naturally, local communities balked at Stanford's blatant arrogance. A Stanford humanities professor suggested that the university should worry less about the biological sciences and begin seeing "as their enemies . . . poverty, fear, ignorance or disease, which should receive the brunt of each [investigator's] attack." A Palo Alto resident chastised Stanford for its "exclusion of the family doctor." And a self-anointed review panel in San Francisco criticized Stanford for placing their want for "gold-plated and super-atomic specialists" over the needs of the community. None other than Windsor Cutting, Stanford's dean of medicine, sympathized with his critics and spoke openly about the agony of his decision to break the agreements.[15]

Terman, however, was noticeably less troubled, and scolded Cutting for equivocating. In Terman's mind, the separation of the biological sciences from the "intellectual and disciplinary limitations of patient-care traditions" constituted a reasonable strategy, and he saw "no reason to involve Palo Alto" in the university's internal affairs again. When the Palo Alto City Council asked for a divisional council to arbitrate their dispute with Stanford, Terman lashed back in the student newspaper: "the primary function of the Medical School is to educate and not to render medical service; excessive service . . . detracts from the research . . . activities of the faculty." Terman's allies rallied, including Stanford Trustee David Packard, who broke from his own paternal managerial style at Hewlett-Packard and instructed Shirley Temple-Black in no uncertain terms that clinical care required "heavy expenditures of capital, . . . something that Stanford cannot afford to do." Sensing that the ground had abruptly been cut from beneath, a handful of administrators and a despondent Dean Cutting either resigned their positions or gradually drifted into estrangement from the biomedical school. Terman merely exchanged them for more sympathetic staff, none more so than Cutting's replacement, Robert Alway, who wholeheartedly embraced the new scientific dimensions of the age.[16]

Dean Alway was no investigator; he was a pediatrician with neither research training nor scientific expertise. However, Alway's early experience as a practicing physician at Stanford taught him a lasting lesson about the university's indifference toward medical care and their subtle but growing interest of pure research. He may have built a career treating sick children, but he was also pragmatic, a consummate teamplayer, and a self-proclaimed "direct and outspoken administrator" who understood quite clearly his charge to push research as the principal focus of the Stanford Medical School. "It must be borne in mind that in contrast to past times medical progress now comes not from the bedside but from the laboratory . . . and more than casual and transitory acquaintance with the basic sciences." "This is not the moving of the old school to the campus," explained Alway to an individual astonished by the sweeping changes underway, "it is the establishment of an entirely new school . . . dedicated to the production of well-trained basic medical men." And sometimes Dean Alway could be brutally straightforward: "The medical faculty's prime reason for existence is research, not practice."[17]

Stanford's purge of those who opposed a research-oriented hospital should not obscure a central truth: in the broadest sense, this crusade was for the soul of the biological sciences. It was neither the first time this battle was fought nor the last. But in the uniquely malleable aftermath of *Sputnik*, with the direction of a rapidly expanding scientific field for the taking, the generosity of science policy provided by its very nature political power, and spelled a messy end, at least for a time, to the public's sincere but unfocused call for patient care. To be sure, those who wanted greater attention paid to medical care also abetted their own decline by refusing to back what many derided as "socialist solutions," such as proposals for a municipal medical insurance plan, medical care subsidies, or strict regulation of the open-market pricing system. Having accepted at face value the principles of state-supported research and free-market medicine, Stanford's opponents defended their cause with passionate pleas for "a more compassionate medical pricing system," a decidedly inadequate strategy considering that the weight of science policy and the power it gave to academia easily overwhelmed such naïveté.[18]

Having dismissed its critics, Stanford turned toward the daunting task of rebuilding a scientific field that had suffered from years of willful neglect, collegial isolation, and limited budgets. At the very least, money for scientific research would not be a problem. It flowed so profusely that even old-guard investigators, once preoccupied with protecting their professional autonomy, began reaching out to their nemesis with grant requests of modest size, typically under $20,000: for instance, Paul Kirkpatrick obtained $16,086 to test mirrors that produce x-ray images for a biophysics project, while $9,436 went to Victor Twitty and $4,212 to Clark Griffin for general research support in biochemistry.[19]

Provoked by Terman to "ask for more," a group of investigators decided to test what they believed were the plausible limits of federal patronage when they submitted an application for millions of dollars to build a biophysics laboratory in the new medical school quad. To later generations, and perhaps to their counterparts in physics and engineering, the amount they wanted might have seemed insignificant, but to contemporaries within the biological sciences, it was enormously ambitious—and wildly successful: the NIH Health Research Facilities Program gave them $1.5 million to cover construction costs for a new biophysics program.[20]

The effect of this award on Stanford's biological scientists was visceral. From the launch of *Sputnik* in 1957 and the opening of the on-campus medical school in fall 1958, Stanford bioscientists received 21 separate awards from the Public Health Service totaling $447,000, all in support of "fundamental research." A single NIH institute gave $943,412 to help investigators establish a radiotherapy program. And individual investigators reaped immediate rewards too, including the aptly named Ronald Grant, who received $16,900 from the Public Health Service, $25,000 from ONR, and $100,000 more from other federal programs to purchase one Tiselius electrophoresis apparatus. Incidentally, in contrast to the extravagantly available federal patronage for research in the biological sciences, the March of Dimes gave $500 to the Stanford "medical student that displayed excellent clinical and bedside skills," a pittance compared to the hundreds of thousands of dollars pouring into bioscience laboratories at the time.[21]

It was a remarkable transformation, and yet too much can be made of federal patronage as the sole reason why research gained such a foothold in the biological sciences at Stanford. Certainly, federal money lent energy to the crusade, but the motivation to conduct expansive pure research went much deeper. For instance, in 1959, medical school administrators departed from bottom-line concerns and added an additional year of classes in basic biological sciences to the traditional four-year medical training program. The move was justified as an attempt to develop the "whole" doctor, but in reality Stanford had taken a significant step toward redefining the meaning of medical education. It also fit perfectly with Stanford's intentions to establish its biological sciences as a leading research program in the country. Four of the top five medical schools thought so too, when each implemented the same changes within the year. At a stroke, Stanford thus earned the admiration of two disparate elements in medicine: bioscience investigators and medical education programs. Stanford had also taken a giant step in the direction of "modernizing" its curriculum and laying the foundations for further advance in the field. Surprisingly forgotten in later years, the five-year plan was arguably one of the forward edges of Stanford's emergence as a leading program in the biological sciences.[22]

Virtually everything accomplished by Stanford thus far consisted of building a core organizational foundation to support pure research in the biological sciences. Yet even as an institution began to take shape, doubts were multiplying about its permanence and its survivability. What Stanford lacked, and desperately needed, was faculty capable of carrying forward their mounting successes. What the university had, and absolutely detested, was a "deadwood" biochemistry department with a staff that would not collaborate with chemists or physicists and would not conduct fundamental biochemical research, as had happened with positive results at other research universities. However, the issue that divided administration from faculty in the biochemistry department the most was federal patronage. These conscience-ridden old-school biochemists categorically refused to subordinate teaching and the training of doctors to fulfill—what department chair Hubert Loring considered—an obsession to conduct research on the public dole.

The inflexibility of Stanford's biochemists, long an obstacle, now galvanized a newly confident administration. Frederick Terman effectively opened Stanford's search for new faculty when he wrote a handful of biochemists flourishing in the heated atmosphere of DNA to ask if Stanford "appeal[ed]" to them. Of these, none seemed at first a more unlikely prophet than Arthur Kornberg, a second-generation Jewish immigrant raised in New York City whose background in medicine, limited training in Europe, disinterest in popular physiochemical studies, and focus on less popular topics like enzymes seemed an improbable base from which to build a career.[23]

Kornberg's atmospheric rise in the profession began inconspicuously with the arrival of DNA's double-helix structure and the new questions about basic biological processes that it raised—in particular, the search for the enzymes that make DNA and RNA. In fact, Kornberg had already set out to test whether enzymes catalyzed nucleic-acid chains when Watson and Crick made their discovery, so he was quite well positioned when experimentalists began to accept that DNA had replaced protein as the "central component, the basic building block of life." In 1955, almost three years after Watson and Crick's discovery, Kornberg discovered polymerase—a naturally occurring family of enzymes responsible for the replication and repair of DNA and RNA. When Terman turned up the recruiting heat in 1957, Kornberg determined that polymerase allowed a specific strand of DNA to act as a template for the replication of itself—without the intervention of any other substances. When Terman decided Kornberg was a "perfect fit," his choice had just used a polymerase enzyme to carry out a series of important and elegantly simple experiments on DNA and RNA, including controlled and unlimited replication, or "synthesis," of specific DNA in vitro. To a great degree, Kornberg helped transform the field of enzymology from one that seemed tainted by medical concerns—in the early years, the field usually attracted investigators trained as M.D.s, much like Kornberg—into a scientific floodgate through which flowed new ideas about fundamental processes and mechanics. In fact, Kornberg's work on polymerase would eventually provide the platform for DNA synthesis, and in a later era would serve as a critical theoretical link that connected the basic research discoveries of his generation to the genetic engineering and cloning experiments of the next.[24]

It was no easy task to lure Kornberg to Stanford. He had taken over Washington University's struggling biochemistry department at the young age of thirty-six and helped turn it into the "mecca of enzymology"; he also knew about the problems that plagued Stanford's attempt to move its hospital to Palo Alto. Then there was the unruly presence of the obstructionists—notably, Hubert Loring—and the challenges that they had apparently caused. But Terman was determined to find something that might bring Kornberg to Stanford, so he asked for a workable wish list that would meet his most unrestrained ambitions. Nothing on that list could be characterized as extravagant—a marginally higher salary, more technical equipment for biochemical analyses, including numerous microscopes and accessories, precision polarimeters for measurements, refractometers for determining substance density, microphotometers, and spectrophotometers—except Kornberg also asked for an appointment that would allow him the unchecked power to create and redirect the direction and temper of a new department with changes to faculty and policy as he alone saw fit. And he also wondered if Terman could "do something about the weakness of the current department."[25]

Terman went further. Backed by the university president and dean of the medical school, and understandably bypassing old-school faculty, Terman matched and in some cases surpassed every single criterion submitted by Kornberg: a raise from $16,000 to $20,000; an increase in the department's annual budget from $80,000 to $200,000; a leading role in the search for new chemistry, biophysics, and biology chairmen; more graduate and postdoctoral fellowships and additional research space for them; "insulat[ion] from 'course-happy' people" in biochemistry; and half the current teaching requirements. It was a tremendous offer, but Terman was not finished: should Kornberg so wish, Stanford would replace the current faculty practicing biochemistry with his entire department at Washington University, including all of his associate and assistant professors, instructors, the secretary, carpenter, janitor, two technicians, "down to the last dishwasher."[26]

It was a masterful recruiting ploy. Kornberg's handpicked staff at Washington University was a rising force in the biological sciences, a composition of like-minded and similarly experienced investigators, organized first and foremost toward the study of biochemical processes at the molecular level. From this base, Kornberg had selected additional staff that ensured an even distribution of scientific expertise and did not overlap. Consequently, he invited only one person per subdisciplinary field to come with him to Stanford, an approach that contrasts sharply with the gargantuan research organization run by Wendell Stanley at Berkeley's BVL: Dale Kaiser had training in virology, so Kornberg did not extend an invitation to his other virologist, Bob DeMars; for physical biochemistry, Kornberg chose Paul Berg over Jerry Hurwitz, primarily because "Jerry [was] more combative and blunt . . . and competitive." Nor did Kornberg attempt to place an investigator in every subfield. For instance, like many in the biological sciences at this time, Kornberg's "aversion to clinical investigation" meant that accomplished medical researchers, such as Irving Loudon, had no chance of securing an appointment in Stanford's select program. Kornberg later admitted that he agonized over the selection process and his decisions caused at least one, and perhaps two, investigators to suffer "nervous breakdowns," but the primacy of experimental success required "internal cohesion," and that meant thoughtful and necessary "external exclusion."[27]

Despite the department's small size and lack of scientific focus, Kornberg appreciated but did not demand that participating investigators conduct research related to his field of enzymology. He did not have to—his own work offered plenty of stimulating experimental opportunity for further study. Certainly Kornberg was not the first to assemble a small and select staff focused toward a singular experimental line, but he may have been more self-conscious about and committed to a cooperative approach than most. That is what associate professor Paul Berg concluded when looking back on his first few years under Kornberg's watchful eye: "Kornberg had a history of directing the work of people in his lab . . . but he allowed me to do what I wanted to do too." Moreover, though Terman's offer to hire Kornberg's entire department was certainly unusual, it reflected a broader effort to promote cooperation that drew heavily on corporate restructuring models. Indeed, Terman probably recruited Kornberg's entire biochemistry department at Washington University as a strategy to protect against the rise of another renegade individualist like Hubert Loring.[28]

The added incentive of bringing Washington University's entire biochemistry department to Stanford placed the offer in an entirely new light. Fired by inspiration, Kornberg accepted Stanford's offer on 5 July 1957; in a stroke, Stanford's biological sciences were transformed. Naturally, research support from the federal government jumped significantly the moment Kornberg and his staff arrived, including but certainly not limited to three PHS grants totaling more than $250,000 and a single National Science Foundation grant for almost $150,000. The experimental equipment under Kornberg at Washington University was a monument to America's advancing technological prowess, and it was all moved to Sanford, sometimes disassembled and shipped piece by piece, lens by lens, pipette by pipette, and even reagent by reagent. But most impressively, Kornberg's entire staff followed, twenty-two people in all—including his administrative secretary, laboratory manager, sculptor, and instrument maker—each displaying a remarkable sense of interest, loyalty, and confidence in what they were about to achieve. And if Stanford administrators wanted Kornberg from the outset to sit at the helm of a magnificent program, his staff would surprise them with their own eagerness and productivity. Investigators such as Cohn, Hogness, Kaiser, and Lehman were all considered top flight by many observers, but it was Paul Berg, then just out of Case Western Reserve's graduate program in biochemistry, whose leaping mind and artful imagination would frequently outpace the others, including one day Kornberg himself. And as for Loring and previous generation of biochemists? Kornberg kept them assigned to the chemistry department, where their "lack of theoretical vigor, strong ties to medicine and their commitment to teaching . . . could not harm" the incoming department's serious research objectives.[29]

Biological scientists from the world over watched with admiration and not without a touch of envy as rumors of Stanford's revival captivated the field and set the stage for an even greater crusade. Joshua Lederberg, a young geneticist at the University of Wisconsin, had grown especially impatient with the way his colleagues practiced genetics research: they saw little value in seeking the visual image of the gene, rarely conducted physiochemical studies, and were hopelessly preoccupied with crosses and maps. Certainly, Lederberg had nothing but praise for their work with vital functions of several "growth factors"—hormones, vitamins, amino acids—and their animal-feeding studies clarified the cause and prevention of certain animal and human diseases. But to Lederberg, geneticists at Wisconsin had overplayed the agricultural aspect of genetics research and had compromised their objectivity by working too closely with the state's burgeoning dairy industry—undeniable applied concerns at a time when pure research reigned supreme.[30]

Isolating himself from his colleagues, Lederberg developed what one historian of science called "a school of molecular genetics almost single-handed." His career actually began in 1946 at the remarkably young age of twenty, when he showed that a "male" strain of *E. coli* bacteria transferred the genetic information that it carried in its nucleic acid directly into a "female" recipient. Pushing this result, Lederberg then showed that two bacterial strains exchange only that genetic information the other needs to become biologically active. These two fundamental discoveries contributed to the then not fully substantiated proposition that nucleic acid—and not protein—carried genetic information; moreover, these two experiments established Lederberg as a leader in his field, primarily because he had directly challenged Wendell Stanley's award-winning work with TMV that "proved" just the opposite. Much like Kornberg's work with polymerase, Lederberg's discoveries also offered the more distant possibility of inserting known genes deliberately into a cell—a process that would, in the next generation, become the basis for genetic recombination and cloning, and the principal technique used by genetic engineers.[31]

In winter 1957, Joshua Lederberg wrote to Frederick Terman and Arthur Kornberg to see if they had an interest in bringing a geneticist on board. In this carefully crafted letter, Lederberg showed a firm grasp of the accepted hierarchy of research and teaching, and the general temper of fundamental research and his belief in its undeniable importance: "Clinical genetics is quite unimportant when compared to the necessity of genetic insight—the function of a medical genetics department is for the education of the faculty more than the students. . . . As for teaching genetics to medical students, I'd want more experience on this point, but frankly this is our least important function. This is the only administrative question I get much steamed up about."[32]

Neither Terman nor Kornberg could ignore Lederberg's overture, nor his passionate commitment to pure research. Indeed, when asked by Terman whether Stanford should add Lederberg to its embryonic biochemistry department, Kornberg paid the applicant an exceptionally exact yet generous compliment. He described Lederberg as a "red hot" investigator and told Terman that he would be a fine addition to Stanford's faculty. But he also told Terman that he did not want Lederberg to join his biochemistry department. Apparently, Lederberg's experimental commitment to fundamental genetics research conformed to Kornberg's highest scientific expectations, but his impatient experimental approach, renegade investigative style, and short scientific attention span did not. Most likely, Kornberg was worried that a geneticist might not be able to establish acceptable distance from agricultural or medical concerns. To a certain extent, Kornberg's assessment of Lederberg's research was on the mark: most molecular genetics research requires laboratory commitments on the order of hours, or days at most; consequently, none of Lederberg's colleagues had ever seen him spend an entire year on a single problem. In this sense, Kornberg concluded that genetics research made Lederberg different, or vice versa. Either way, Kornberg believed that Lederberg's experimental habits, or his experimental focus, threatened the professional dedication of Stanford's nascent biochemistry department to fundamental research.[33]

Terman, characteristically driven to great lengths to stimulate Stanford's biological sciences and blessed with a windfall of federal money now available in the post-Sputnik era, obliged Kornberg's stoic provincialism and carefully presented Lederberg with an offer to establish an entirely new department of genetics in which he could be, if he so wished, the sole member. Lederberg, perhaps unaware of Kornberg's reluctant interest in him, accepted the offer and ironically arrived at Stanford in 1958, a few months before Kornberg and his biochemistry entourage from Washington University.[34]

Securing Arthur Kornberg and Joshua Lederberg, perhaps more than any other factor, solidified Stanford's elite place in the biological sciences. Almost immediately, applications to enter Stanford's graduate programs in the biological sciences grew by 20 percent. Federal patrons of science also showed an even greater surge of interest to sponsor research at Stanford. And then, quite unexpectedly, two welcome events further brightened the picture: amid the chaos of recruitment, staffing, construction, and relocation, both Joshua Lederberg and Arthur Kornberg received the Nobel Prize for their respective work. With two laureates in a small program, notoriety became almost inevitable and gave Stanford, in the words of Dean Alway, "a medical school . . . praised by our enemies rather than defended by our Alumni."[35]

Certainly no one could deny that the biological sciences had improved considerably at Stanford. However, the situation was far from perfect. In a confidential year-end review, Dean Alway presented hospital administrators with an overwhelming list of twenty-seven serious problems that he believed could disrupt the medical school's future; most were holdovers from the move of the hospital and separation from the Palo Alto agreement. The community's concerns about clinical care remained unanswered too. Overshadowed by these more pressing issues, the possible incompatibility of their two brightest stars might prevent profitable interdisciplinary collaboration between two related fields—genetics and biochemistry—but this matter hardly registered on Stanford's collective radar screen.[36]

The Rise of Basic Bioscience Research at UCSF

Across the bay from Berkeley and up the coast from Stanford lies San Francisco, where, beneath the early morning and late afternoon fog in the city's western-most district, sits the University of California, San Francisco Medical Center (UCSF)—known then, formally, as the University of California, College of Medicine. UCSF is perched on the city's side of Mt. Parnassus. The hospital's location, on a rare clear day, is one of the most picturesque academic settings in the world; the university looks down upon virtually every scenic San Francisco site, including Coit Tower, the San Francisco skyline, Golden Gate Park and Bridge, and Mt. Tamalpais across the mouth of the Bay.

Its charming setting clashes cruelly with one of the most formative natural catastrophes in human history. On 18 April 1906, at 5:12 A.M., a violent earthquake struck the heart of San Francisco. In few places did the San Francisco earthquake cause more damage than at UCSF, which was partially buried by the rubble raining down from Mt. Parnassus and further destroyed by a fire that burned for days. Eyewitness accounts of the conditions at UCSF paint a horribly gruesome scene: "Operating tables were filled all the time"; "infants brought in in their mothers arms, were laid down, despite suffering from mortal burns [*sic*]"; "many men and women, caught by falling walls and horribly mangled, in many cases broken bones protruding through the flesh, nevertheless had to wait while more serious cases were attended [*sic*]."[37]

In a remarkable act of generosity, Berkeley administration offered to take in UCSF's non-essential preclinical staff—in particular, professors of anatomy, physiology, and biochemistry—so newly vacated offices and classrooms could be used to treat the incoming wounded. Out of habit and over time, UCSF became known as a clinical center, a designation that stood in stark contrast to the reputation of its academic brethren, Berkeley, as a research university located directly across the bay. The simple divide of pure research at Berkeley and clinical training at UCSF owed much of its durability in the postwar period to the straightforward administrative structures implemented as desperate responses to the 1906 San Francisco earthquake.

In the post–World War II era, the dramatic attempt by Berkeley to establish a dominant program in the biological sciences, followed by Stanford's sudden entrance into the field, put UCSF at a crossroads. The biological sciences had consistently lain beyond the grasp of UCSF's small staff of investigators because an overwhelming number of practicing physicians held preponderant political, financial, and organizational power. The opportunity to do pure bioscience research had receded even further when Berkeley began its aggressive campaign to build the

BVL; the California state legislature and the University of California Regents failed to see why the state needed two bioscience research facilities. Moreover, Stanford's decision to move its hospital from San Francisco to Palo Alto left UCSF as a critical provider of medical care for the city's exploding population.

But in the aftermath of *Sputnik,* a number of elements essential to establishing a bioscience research program at UCSF began falling into place. About the time that *Sputnik* was orbiting the Earth in 1957, UCSF officials celebrated the addition of the thirteenth, fourteenth, and fifteenth floors to the Moffitt Medical Tower. Scarcely a week after *Sputnik,* the California state legislature agreed to match the federal Congressional Health Research Facilities award of $460,000 to help UCSF pay for additions to the new medical tower and promised to provide additional support for two more medical towers in the near future. To add point to the momentum that had swung in favor of those few investigators pressing for a research program at UCSF, the NIH promptly responded to the federal and state awards with an additional grant of $230,000 to support construction costs, an award that also signaled the institutes' new interest in helping to establish pure research in medical schools.[38]

But faculty in the biological sciences at UCSF worried that this handful of policies, unapologetic availability of state money, and eager promotion of change provided only opportunity and no assurances that anyone in administration would support them in their efforts to overcome the medical school's unyielding clinical staff. They worried even more about the future when the regents named Clark Kerr, who had an economics Ph.D., to succeed Robert G. Sproul as UC's next president, rather than someone who had made his mark in a scientific field.

Clark Kerr had made a career out of unexpected promotions. Fresh out of graduate school, Kerr dared to accept a position as a labor arbitrator on the War Labor Board, and then promptly convinced an angry labor union to settle its dispute with management in a Seattle munitions plant. Soon after, UC regents invited Kerr to come to Berkeley and mediate ongoing loyalty oath controversies, which he did with great success. Faced with swelling enrollments and a pressing need to decentralize administrative authority by delegating power to individual campuses in 1952, UC Regents again turned to Kerr and asked him to serve as the inaugural chancellor for Berkeley, where he remained for almost five years despite languishing under President Sproul's overbearing micromanagerial administrative style. In this newly created post, Kerr made important work out of apparently dead-end jobs to build an impressive administrative resume: his consistent deferral of authority to department chairs showed his respect for autonomy and decentralization; his approval of virtually all applications for federal patronage that crossed his desk established him as a sympathizer of pure scientific research. Later, Kerr won accolades for casting fundamental research as "one of the 'multiversity's' primary functions in society" in his draft of California's Master Plan for Higher Education. Bioscientists in the UC system eventually came to see Kerr's ability to delegate and mediate, and his unwavering support of pure research, as sure signs of a strong leader, or so it seemed at the time.[39]

First as chancellor, then as president of the University of California, Kerr devoted much of his energy toward advancing the UC's academic reputation. Kerr was not content with the acclaim that the University of California was ranked the "second best balanced distinguished university," behind only Harvard. To Kerr, the UC's top-five ranking in twenty-seven of twenty-eight departments made it easy to identify the lagging field: it was UCSF's lean research output that stood out. To confirm the results of the poll, Kerr and his associates contacted medical school deans from around the country and asked them, in essence, to rank their own field; most agreed that because UCSF did not have a vibrant basic bioscience research program, it was only the twentieth to twenty-fifth best medical school in the country. Meanwhile, a similar study conducted by the American Association of Medical Colleges stated that UCSF's resources "outstrip the quality of the faculty and the program," an assessment that added further insult to injury.[40]

A deeply troubled Kerr found the evaluations "intolerable," and immediately made several key appointments in terms of both research policy and academic politics that he hoped would offset the physicians' "proprietary force" at UCSF and push through a more productive basic bioscience research program. None of the appointments proved as crucial as Julius Comroe to head the newly created Cardiovascular Research Institute (CVRI), and John Saunders as dean of medicine. To many veteran UCSF staff, both Comroe and Saunders had the ability and energy to turn UCSF into a first-rate bioscience research center; however, UCSF's atmosphere of perpetual disciplinary ferment, the incompatibility of their visions, the strong-arm tactics each man used to implement them, and their autocratic personalities nearly destroyed UCSF's best chance to build a basic bioscience research program.[41]

It was in spring 1956 when Kerr first contacted Comroe about his interest in running a dedicated bioscience research program at UCSF. After ten nondescript years as an associate professor at the University of Pennsylvania, Comroe concluded that "East Coast traditions" stifled great scientific potential and declared himself ready to make a change. A proud man in search of professional distinction, Comroe visited UCSF with high hopes, but came away unimpressed. So, too, Kerr. Comroe believed that the physicians at UCSF wielded too much power; Kerr sensed that Comroe was a "very able guy, but something of a wild-man . . . a very aggressive, perhaps even irresponsible person." But Kerr recognized that Comroe's passion and determination would be a tremendous asset when it came time to do battle with the university's reigning medical establishment and forcibly establish a basic bioscience research program. With a few misgivings, Kerr nevertheless offered Comroe the chairmanship of a wide selection of departments, including pharmacology and pharmacology-physiology; each time Comroe refused. Kerr then suggested that he could serve as director of UCSF's new CVRI, even though he had no formal training as a cardiovascular physiologist. Comroe accepted the offer, but only on the condition that he did not have to report to the university's parochial medical review board, which was naturally dominated at the time by a large number of practicing physicians.[42]

"What the hell [is] going on in San Francisco?" asked a friend and former colleague of Comroe when he heard about the new research programs at UCSF. He should have known. The moment Comroe arrived, he marched the CVRI headlong into basic bioscience research with a major promotional drive to attract recruits and federal funding, bypassing in many instances the "snail-like decision-making processes" that characterized most UC appointments. Comroe tapped some of the most prestigious investigators in the country, some of whom, like Manuel Morales, shared a similar "disenchantment" with East Coast traditions. He also plumbed the ranks of local hospitals in search of physicians who had become frustrated with medical practice and who also showed a talent for and interest in cardiovascular research, including Isidore Edelman, the chief of medicine at San Francisco General Hospital, and Elliot Rapaport, who was chief of the cardiac center at Mt. Zion Hospital. Comroe would later describe his early recruiting successes as "the reverse domino theory," in which the recruitment of "one highly respected scientist attracted more." To increase laboratory space for the swelling staff, Comroe played on the triskaidekaphobia of the medical establishment by promising recruits valuable research space on or near the thirteenth floor of the newly expanded Moffitt Medical Tower. Comroe's recruitments and laboratory expansion efforts fueled an incredible surge in research support, growing from a dismal $105,159 to almost $1 million, including an immediate windfall of $220,000 when Comroe and Morales agreed to transfer from Penn to UCSF.[43]

Fired by newfound inspiration, UCSF's student newspaper, the *Synapse*, praised the staff at the CVRI for turning the university into an "ivory-tower" and waxed admiringly about the school's newfound legitimacy. Clark Kerr also appreciated the CVRI's ability to sustain its burgeoning research programs and encouraged investigators in other research units to seek more research support from federal agencies. In part because of the success of the CVRI, the normally placid UC Regents contributed to UCSF's newfound independence by designating the university as a separate campus from Berkeley. Most illustrative of the enormity of the changes taking place, UCSF physicians, though still the majority on campus, were beginning to soften their opposition and even, in some cases, publicly supported the new direction.[44]

As it turned out, the independence that Comroe used to engineer a remarkable turnaround at the CVRI was short-lived, when shortly afterward Kerr appointed John B. deC. M. Saunders as dean of the medical school—Comroe's new and only supervisor. Unlike Comroe, whose background and training in the United States had instilled in him a deep appreciation of the separateness of fundamental and applied research, Saunders grew up in Johannesburg, South Africa, and then studied medicine at the University of Edinburgh, where he learned to appreciate the unique Scottish blend of research and clinical care. After completing all of his training overseas, Saunders became a member of the anatomy department at Berkeley; his arrival coincided with the arrival of Wendell Stanley, and he witnessed firsthand the "curiously American approach to medicine" that split research from clinical care. He was not terribly impressed. Saunders's constant defense of "balance" between research and clinical care provided him, paradoxically enough, a safe distance from the university's lacerating internecine battles. He even turned his mediating position into an advantage when he presented to the administration a comprehensive plan to improve America's health-care system, one in which the biosciences, medicine, and the humanities all converged around the needs of the patient—a program that he referred to as "human ecology." His vision was thoroughly unconventional at the time, and he took pride in it, promoting himself and his plan in such vainglorious fashion that colleagues often joked about which of UCSF's two newest figureheads—Director Comroe or Dean Saunders—had the most inflated sense of self.[45]

As novel as Saunders's vision for improving medical care was, Kerr selected him as the new dean of medicine for other reasons too. Kerr was "impressed" by the South African's sophisticated persona—he would admit in his memoirs that he often thought of Saunders as "royalty"—and, much like his assessment of Comroe, believed that such a strong personality would come in handy when the rest of UCSF followed the lead of the CVRI and shifted its focus away from patient care and toward research. But most of all, Kerr appreciated Saunders's "enlightened" goal of elevating basic bioscience research to improve UCSF, choosing to ignore of course Saunders's equal emphasis on balancing research with patient care. Kerr was not the only one mesmerized by

Saunders. Most everyone who came in contact with him appreciated his carefully crafted image as a visionary. Nor was Kerr the only one to misinterpret Saunders's commitment to pure research: most investigators at UCSF mistakenly assumed that research was the primary leg of Saunders's "human ecology" triangle, when in fact the author merely saw it as UCSF's weakest link in temporary need of propping up in order to create a more balanced medical program.[46]

From the outset, Saunders relished the power that came with his appointment as dean of medicine and wielded it with gusto. He immediately set out to build a stronger basic bioscience research staff by dividing to conquer. He began by inviting to UCSF his fellow preclinical investigators at Berkeley who had suffered under Wendell Stanley's disciplinary purges. Of course, a vast majority of Berkeley bioscientists refused, including old-school investigators such as the biochemist Howard Schachman, who did not believe UCSF "had it in them" to make the full transition; Hubert Evans, who continued to disparage the medical school as a "hopeless trade school"; and the endocrinologist Choh Hao Li, who expressed interest but nevertheless required direct access to Berkeley's elaborate chemistry laboratories. However, a handful of Stanley's minions jumped at the chance to escape the debacle of the BVL and start anew at UCSF, such as Berkeley's entire physiology department, which included first-rate investigators such as Ralph Kellogg, Peter Forsham, and Richard Havel.[47]

Whatever the scientific merits of accepting a transfer from the "second best research university in the country," Saunders's calculating offer cut directly to the heart of UCSF's other "peculiar problem." The lack of research productivity, Saunders believed, was due to the fact that the university was "entirely deprived of research support"—a plausible opinion since federal patrons awarded research output with more research support, creating a self-perpetuating system in which the rich indeed got richer. The arrival of transfers from Berkeley, Saunders had cleverly realized, would bring much needed patronage and equipment, especially in the aftermath of *Sputnik,* and might even establish UCSF as a "federal grant university." His plan worked as well as anyone could have imagined. Consider, for instance, the profound transformation that occurred when Leslie Bennett moved his metabolic research unit from Berkeley to UCSF in 1958, which increased the medical school's entire anatomy department budget from $117,561—by ordinary measures a pitiful sum even then—to well over $250,000.[48]

More effective than Comroe's individual efforts in the CVRI, Saunders's early successes as dean unleashed a broad base of pent-up ambition to conduct basic bioscience research at UCSF. Investigators looked to Saunders as their resonate soulmate, inundating him with a host of new research proposals: was he not one of them, a sympathizer who originally came from an academic department at Berkeley and who supported basic research as they did? From all sides, pressures played on Saunders to commit to this or that basic research program. His balanced posture in these early days, along with his immediate attempt to elevate basic research guaranteed a wild ride, marked by a desperate fervor with which investigators would urge more.

Saunders' gentle push to improve the standing of basic research in 1958 became, in less than two years, a powerful and independent force. The surge of interest in basic research that had seemingly enveloped UCSF stunned Saunders, who recognized that Comroe and his associates at the CVRI were leading the headlong march without challenge. Indeed, Comroe pressured David Wood, Director of the Cancer Research Institute, into giving up space on twelfth floor at Moffitt Hospital to Comroe's CVRI; a clinical care unit on the fourteenth floor also transferred space to the CVRI, space that once held sixteen beds for patients and a nurses' station. Comroe flatly refused to allow his staff to teach lecture courses, arguing that such time commitments would reduce time spent in the lab, and encouraged other investigators to do the same; staff in the CVRI pressured long-time biochemistry chairman and a loyal advocate of patient care, Max Marshall, to retire, and then quietly attempted to replace him with one of their sympathizers. Adding insult to injury, Saunders received bills totaling $75,000 for unauthorized equipment taken by investigators who had transferred to the CVRI from other research universities, including calculators and simple experimental equipment.[49]

To Saunders, the tide of staff sentiment, federal support of scientific research, and administrative policy all seemed to be bowing subserviently before Comroe and his push to establish a basic bioscience research program, but as dean of the medical school he had the power to reestablish medical care on an equal footing. The counterattack that Saunders mounted was nothing less than impressive. He reprimanded UCSF investigators for their neglect of the patient and their haste to conduct fundamental research. He tabled all grant applications for at least one month to force applicants to think carefully about their proposal before submitting it to committee for peer review, a move that established UCSF as "the most administratively inefficient research university in the country," according to one NIH report. He vetoed UCSF's carefully planned and well-funded effort to establish a School for Basic Research, arguing that "a separate Fifth School dedicated to research [was] counterproductive," especially at a medical school committed to balancing research and medical care, and then promptly sent the proposal to the UCSF Academic Senate where a majority of physicians heeded Saunders's recommendation and summarily voted against it. He also redirected a large donation intended for the CVRI to the university's general account where it could be divided among all staff more equitably. Finally, Saunders rejected Comroe's nomination for a new biochemistry chair.[50]

If anything, Saunders had been so effective at reining in basic bioscience research at UCSF that he no longer had to worry about the strength of the investigators' new commitment to research, but the strength of the counterattack that his own countermeasures provoked. Fighting Saunders, not finding new bioscience knowledge, became the principal task of the research staff at UCSF who interpreted Saunders's actions as a direct, frontal assault on a program that had just gotten started, especially so for Comroe, whose own elevated sense of basic research clashed mightily with Saunders's vision of balance. One way to overcome Saunders's stubborn authority, reasoned Comroe, was to overwhelm him with grant applications, showing, in effect, the staff's dogged devotion to research. The move backfired, merely antagonizing an already beset Saunders. In an extraordinarily brazen memo, marked by curt frustration and condescension, Saunders responded to Comroe's repeated requests for money and the futility of the continuing charade:

Memo re: Comroe:

1. On 11/1/59 [you requested] $20,000 to purchase needed equipment not previously funded.
2. On 11/10, $26,000.
3. On 11/22, $10,000, increased to $16,000.
4. On 11/24, money . . . for contingency—$10,000.

When will it stop?

—JBC Saunders[51]

Then, on 8 June 1961, Saunders went too far. In an attempt to curtail the growing influence of investigators committed to basic research, Saunders unilaterally forced UCSF back into the business of patient care by quietly signing a cooperative agreement with Franklin Hospital in San Francisco (SFFH) and then turned over a tremendous amount of space at UCSF to SFFH clinical practitioners and the authority to make many joint appointments. The proposal to merge SFFH with UCSF, if passed, would have nearly doubled the number of physicians on the UCSF staff. Investigators such as Comroe reacted with outrage, and even a great number of the physicians wondered aloud about Saunders's sovereign decision and the wisdom of adding still more clinical staff. Afraid that he might lose the chance to plunge the proverbial dagger into basic bioscience research, Saunders declared in no uncertain terms that he had the authority to override any decision made by the faculty: "since the

Academic Senate is empowered to deal only with academic matters and since the educational policy (of balance between clinical care and research) was set . . . years ago; therefore, the [merger] with Franklin . . . does not concern the Academic Senate." Given the decades-long agitation between physicians and investigators at UCSF and the persistent attempts by a minority of investigators to overpower the physicians' powerful voting bloc, Saunders's decision to bypass the democratic process at UCSF in order to increase clinical care stood out not for its wisdom but for its provocation.[52]

For Saunders's critics, his unilateral acquisition of SFFH was the final insult. Those pressing for basic research vowed they would no longer be so easily dismissed; what Saunders had called medical reform his critics assailed as a "mistake." A fatal opinion took over investigators at UCSF that neither Saunders, nor the physicians, nor the patient would stand between them and their inevitable successes. Their perception of Saunders as the enemy drips with irony: to them, Saunders was the sole obstruction to establishing basic bioscience research at UCSF at the same time that they rejected medical research, an opinion that looked upon different kinds of research as mutually exclusive.[53]

On November 27, 1964, twelve angry investigators carried a petition to Kerr asking for Saunders's removal. Caught up in the tumultuous student protests on the Berkeley campus, Kerr deflected the controversy to Vice President Harry Wellman and asked him to address their complaints. Behind closed doors, the "committee of twelve" presented Wellman with a devastating litany of protests, well documented and substantiated with evidence that showed Saunders had indeed abused his office: Saunders, in their minds, sought dictatorial control over UCSF, which had prevented investigators—and the university—from reaching their potential. They threatened Kerr with an ultimatum: either fire Saunders, or they would quit.[54]

Far from a mundane attempt to replace a struggling administrator, local newspapers broke the story as a "bitter, disruptive struggle on the UCSF campus." Details of the conflict provoked a malicious counterattack from staff physicians, UC alumni, and the local community. In support of Saunders, many threatened resignation and credited his vision of "human ecology" as the first credible attempt by a UCSF administrator to unite research and patient care—no small feat in a politicized city such as San Francisco. Those defending Saunders also responded to the petition signed by twelve UCSF investigators with one of their own, but theirs contained nearly one thousand signatures in support of Saunders and his work as dean of medicine. Internal turmoil marked the political struggle at UCSF, just as it had at Stanford during the move of the hospital from San Francisco to Palo Alto, but it was a state senator who recognized a latent force that would one day have more influence on the future direction of bioscience research than any parochial fight or scientific discovery: "many Californians are getting fed up with attempts of [UCSF] scientist tails trying to wag the university dog."[55]

On 1 October 1965, during the gloomy twilight of his presidency, Kerr instructed Wellman to remove Saunders as dean of the medical school. The full story of the coup that overthrew Saunders and clinical authority at UCSF is rich in theatrics, headlines, and mystery. Its leaders were colorful and emotive characters, such as Comroe and Morales. Of the two petitions, the one that carried more weight was the one with only twelve signatures delivered with conceit by dignified, professorial scientists such as Bert Dunphy and Holly Smith, chief of medicine, and that counted among its supporters a handful of pure research fanatics and narrow preclinical researchers such as Ernest Jawetz, Maurice Sokolow, and Alex Margulis. It contained melodramatic moments that involved William Reinhardt, who attended the closed-door meeting with Vice President Wellman aware that he was the administration's first choice to replace Saunders. It also encompassed the tragic tale of Saunders, promoted ostensibly to rescue UCSF from its dated emphasis on patient care and who capably strengthened research with an affirmation from Kerr, only to have his charge abruptly cut out from beneath him with an impersonal letter from Kerr that Wellman delivered that astoundingly used bureaucratic protocol to inform him of his dismissal. The rupture of the relationship between Saunders and twelve angry researchers at UCSF marked the beginning of the end for Saunders and his meteoric rise in medicine, and he gradually disappeared from UC administration, offering only occasional acrid criticism of UCSF's neglect of patient care.[56]

The leaders of the coup may have found comfort in seeing their vanquished opponent relieved of his duties, but they could not ignore the fact that the university still lacked a robust bioscience research program. And then there was the constant irritation of sick patients requiring treatment and a staff determined to treat them. It was a confusing array of pressures that beset the newly empowered investigators at UCSF; yet, they could not help but think that Saunders's removal had cleared the way for a new era. Comroe described it as nothing less than the sole reason why "recruitment became a joy again," and why the "faculty became both productive and happy." Falsely thinking the path to basic bioscience research now lay wide open before them, investigators set out to build new basic research programs with immediate and effective selections. They may have sensed momentous potential, but they failed to see that internecine competition with the physicians, and more significantly, growing public frustrations, would become a torrent of protest and a much greater threat than previously imagined.[57]

Building Basic Bioscience Research at Berkeley, Again

As noted earlier, in the late 1940s, Berkeley was the first research university in the Bay Area to establish a landmark bioscience research program. Its BVL attracted many leading investigators, including Wendell Stanley, a Nobel Prize winner, to direct the program and all research projects in it. The BVL also secured some of the most advanced experimental technologies available and took advantage of Berkeley's renown in physics by incorporating their remarkable radioactive isotope program into bioscience experiments. In the process, the BVL attracted much interest on campus and throughout the bioscience community, and investigators clamored to participate in this phenomenal program. The staff assembled was a talented and driven group of investigators bound together not just by the accident of timing. Although they represented a broad range of scientific expertise, they shared certain core beliefs: a deep suspicion of applied research and unwavering confidence in the so-called harder sciences such as physics and chemistry. And at one point it looked as if Stanley's BVL would become the foremost protein-centered research facility in the world, leading one observer to gush, not without reason, that "Dr. S[tanley]'s show will throw [all other bioscience projects at Berkeley] strongly in the shade both figuratively and literally, and should make the University of California pre-eminent not only in physics but in biochemistry as well."[58]

Indeed, their agenda had been from the outset something more than simply establishing a bioscience program. They wanted to become a leader, a program that others looked to as an authoritative voice in a confusing field. They also aimed to establish a workable, productive system that could carry on when Stanley retired, one that would not miss future bioscience opportunities. Toward those goals they had made some notable progress, not least the permanent establishment of the BVL. But events had shown that they achieved none of their highest aspirations, nor were any of their larger purposes fully gained. For the discovery of the DNA double helix had torn away at static disciplinary assumptions, causing both the BVL and Wendell Stanley's reputations to come crashing down.

On 24 January 1962, a desperate memo circulated throughout Berkeley administration. Apparently one of their Nobel Prize winners in physics, Donald Glaser, wanted to shift into the biological sciences but had become so disenchanted with the university's inability to establish a leading research program that he was at that very moment interviewing for a position at MIT. Newly appointed chancellor Edward Strong reacted with haste, sending one of Glaser's respected colleagues, Glenn Seaborg, to take the first plane to Boston to try and coax their scientific star back to Berkeley. Strong then placed a desperate telephone call to Glaser himself to find out if the rumors that he had heard were true. Glaser acknowledged that he was indeed trying to decide whether he should stay and wait for Berkeley to sort things out or go somewhere else, to the East Coast perhaps, where scientific patterns and traditions were more established. When asked what it would take to convince him to stay, Glaser replied petulantly that the "status of biophysics . . . is going to be the key to [my] decision."[59]

To many within UC administration, Glaser's threat to leave confirmed that the worst possible scenario for Berkeley had finally come true: their struggles in the biosciences had begun to tear apart physics, a discipline in which Berkeley had always dominated. Seaborg returned a few days later from his trip to Boston and, in a private meeting with President Kerr and Chancellor Strong, acknowledged that Glaser was serious about his ultimatum and warned that "if we lose this Nobel Prize winner, it will be a 'crack in the armor' which will or may set off a 'merry ride of other possible losses,'" including perhaps his own defection. With the UCSF coup still resonating in the back of his mind, Kerr made no secret that the university must "not lose Glaser"—who knew when or where their next defection might emerge? Strong agreed, pointing out with rhetorical sharpness that "money was not so important" as the future of Berkeley. In the end, all three men agreed that they had little choice in the matter and gave Glaser's demands "the highest priority for assignment," more attention than any other issue, including the developing student protests that would ironically have their own dramatic impact on this and the future of the biological sciences.[60]

At various times throughout the course of the next few months, nearly every Berkeley administrator approached Glaser's looming transfer as if the university's reputation ultimately hinged upon piecing together a plan that met the laureate's demands. Glaser's ultimatum touched off an acrimonious, prolonged, and in the end maddeningly inconsequential attempt to establish another leading bioscience program at Berkeley. Not since the BVL had so much intellectual and administrative energy been expended with such meager results. Yet the peculiar array of explanations and interests that contended in this episode, and the particular equilibrium in which it finally came to rest, reveals much about the pressure to conduct pure bioscience research. It also reflects the overriding disciplinary objectives that ultimately would have to contend with the force of coming popular protest.

Berkeley had to decide, as Kerr and Strong understood it, on the proper balance of bioscience fields that might collaborate in a unified biophysics. But Kerr and Strong also agreed that "we would not want such a department to be narrowly conceived." Fatefully, both men fell for old habits, compounded even further when they deferred to the physical sciences' mass and muscle by promising that they could fill four positions in a new molecular biology department with appointments at the assistant professor level. They also assured Glaser that he could select his own appointments, such as the eminent physicist Arne Engstrom from Sweden to fill a newly created position in medical physics. As the molecular biology department began to grow, it took on a direction and tone that resembled Stanley's BVL, leading to fundamental questions about "relating the physical and chemical concepts and techniques to biological material and problems." Ironically, considering Wendell Stanley's poor administrative track record, his endorsement of molecular biology and his offer to replace his BVL "with the new department" should have signaled that the proposal had serious flaws.[61]

Sure enough, Berkeley staff responded with characteristic outrage when they read a preliminary report in the *Cal Daily* that the administration was prepared to establish a molecular biology department without faculty input. Most objected once again to the long-held premise at Berkeley that the emphasis on physics in the biological sciences was just another instance of preferential treatment and demanded that a program in molecular biology instead emphasize "biology [over] the physical sciences."[62]

Faculty protest intensified in mid-November 1962, when an embattled Chancellor Strong received two letters, one from Michael Lerner in genetics and another from two professors in agricultural chemistry, both of which warned that they would not allow their programs to be "forgotten." The next day he received a scathing letter from biochemist Melvin Calvin that reiterated a similar theme. Then came yet another threat from Leonard Machlis that "a program without botany would only complicate the situation." Five more letters arrived the next day, including a group of letters from the bacteriology and immunology departments; they felt that their clinical contributions had been categorically dismissed. All told, investigators from these and various other departments, including cell physiology, nutritional science, plant pathology, and soil and plant nutrition, complained about the special treatment physical scientists received at Berkeley and warned about the possible ramifications that such emphasis would have on this and all other plans for bioscience research on campus.[63]

The least conspicuous critique heard in these weeks was that of the Assistant Vice-Chancellor Alden Miller, who never directly attacked the overall objective of establishing a basic bioscience research program at

Berkeley, but simply the manner in which it was pursued. Miller especially opposed the administration's decision to hire more investigators from outside the university, which he believed already had the potential to build a bioscience program of "great distinction" from within; all the administration had to do was "properly regroup . . . such stars as Stanley, Robley Williams, Calvin . . . Stent and indeed several others." But most of all, Miller took exception to the willingness of UC administrators to kowtow to Glaser and create a molecular biology program, despite the fact that he was trained and built an illustrious career as a high-energy particle physicist. "We must realize," cautioned Miller, "that Glaser, having won his award [the Nobel Prize] in physics, has declared his intention to enter what is for him a new field. . . . I think Glaser should certainly be encouraged to exert his talent in this direction, but we must preserve a sense of proportion and balance, and realize that he is a newcomer to this field."[64]

To his credit, Miller continued to voice his disapproval throughout the debate, reminding anyone who cared to listen that "we should not overestimate G[laser]'s powers or potential as a molecular biologist and that he is not indispensable in this area." But the supremacy of physics and its relation to pure bioscience research was about the only policy upon which the administration could agree, and even then only with some qualifications. In an atmosphere that cast pure bioscience as the penultimate objective of research, compounded by the general tumult that was beginning to challenge Berkeley in the early 1960s, Miller was the only administrator at the time willing to step forward and chastise his colleagues for caving to the lure of building a program dedicated entirely to pure bioscience research.[65]

"It would be disastrous," concluded the chairman of the Committee on Educational Policy, Davis McEntire, to limit the scope of molecular biology in such a way that would "allow any fundamental split to occur." Indeed, McEntire had noticed that the "rigidity of departmental lines" had been the source of many of Berkeley's bioscience problems in the past. Therefore, he and his committee unanimously recommended nothing less than extending an open invitation to everyone who had any interest in molecular biology, including "several basic biological departments within the College of Letters and Sciences (Biochemistry, Biophysics, Botany, Physiology, Virology and Zoology) [and] softer biological departments such as Entomology, Soils and Plant Nutrition, Plant Pathology, and Genetics in the College of Agriculture." Kerr unequivocally supported the committee's decision and insisted with uncharacteristic clarity that he would not allow differences of opinion between the various subdisciplines to "stand in the way of this proposal."[66]

If there was any chance for the kind of thoughtful reflection recommended by Alden Miller, it quickly disappeared when the regents secured a "large loan" from the state of California and grants from the NSF and the NIH to create a new program in molecular biology with enough staff and technology to fill a 300,000-square-foot building. Still, Miller argued against taking the money, pointing out that Berkeley had ample evidence that "one great unified department is always hopelessly unwieldy," but he was still hemmed in by the dreamers and legacies, scarcely able to wield the authority necessary to overcome the chorus of vocal proponents and boldly repudiate the dogma of pure research. Not surprisingly, by the end of 1963, it was Alden Miller, not the supporters of the new molecular biology department, in full retreat; Miller was eventually replaced by someone whose views coincided with the vision of those determined to create a grand basic bioscience research program at Berkeley.[67]

During a University of California ceremony in 1958, Karl Mayer, a longtime microbiologist at UCSF and at one time its leading scientific figure, tried to deepen his colleagues' sudden passion for pure research by pleading with them to commit to health care. "Responsibility" was the heart of the matter to Dr. Mayer. "How might we fail if we focus on basic research?" Mayer admonished. "We have indeed gone through a tremendous shift in understanding the major causes of disease," he reasoned, "but never forget, today, mental illness, nutritional disorders, radiation hazards, traffic accidents, and lung cancer, still provide us with our greatest challenges."[68]

The biosciences had certainly gone through a "tremendous shift," yet by the time Mayer issued his challenge, most investigators at Berkeley, Stanford, and UCSF had already dismissed applied research as hopelessly less worthy. In the relatively short period of time between the end of World War II and Mayer's proclamation, local investigators had helped decipher the genetic code, unravel the cellular machinery responsible for the replication of DNA, describe protein synthesis in considerable detail, and link the structural and biochemical characteristics of DNA to classical problems in genetics. Their research successes helped lay a scientific foundation for others to study natural genetic recombinations, enzymes responsible for expressing DNA, and the regulation of protein synthesis in bacteria. In fact, when Mayer issued his challenge, bioscientists at Stanford were engaged in nothing less than the remarkable attempt to replicate viral DNA in the test tube, with the further dream of synthetically creating the first biologically active gene. "No one can ignore the biological sciences," pronounced an inspired graduate student in Stanford's biochemistry department. "This is, in fact, essentially a revolutionary period of discovery."[69]

Whether the biosciences were revolutionary, most investigators concurred at the time that they were involved in a scientific revolution. They pointed to the incredible surge of fundamental discoveries as evidence of a sharp and permanent change in the direction of the biosciences. But when did fundamental bioscience research become mainstream? It could very well have been directed toward what some refer to as "alternative scientific realities." For instance, investigators could have redirected their attention toward complex cellular systems that had more relevance to the human condition, a topic that would become extremely fashionable in the next decade. They also could have explored the relationship between organisms and their environment or "upward causation"—the study of life and the living. Even certain technical bioscience topics that would become extremely popular for the next generation of investigators, such as the deliberate search to control and manipulate cells at the molecular level, had virtually no play in Bay Area bioscience laboratories during the 1950s and early 1960s.[70]

That investigators would interpret the "tremendous shift" in basic bioscience research as "revolutionary" was far from obvious in the early postwar period. Then it appeared that bioscientists and administrators at Stanford and UCSF hoped to establish modest programs, make occasional contributions to the larger field, keep up with expanding student enrollments, and perhaps attract enough federal patronage to become a self-sustaining enterprise. But something significant happened between the discovery of the DNA double helix and the Soviets' successful launch of *Sputnik.* Indeed, during this latter period, Berkeley built the BVL, Stanford had moved its hospital to the Palo Alto campus, and key personnel had been hired and transferred from Berkeley to UCSF. Certainly, the scientific foundation established by Watson and Crick and generous federal policy for scientific research contributed to the revolution in the biological sciences. But from a broader perspective it appears that decisions made earlier established fundamental bioscience research as exceptional and unique.

The growing importance placed on basic bioscience research can be traced to the early postwar period and came largely from the investigators themselves. Particularly important were those committed to pure research because they were the ones who had enough authority to exclude applied research, such as Wendell Stanley at Berkeley, Arthur Kornberg at Stanford, and Julius Comroe at UCSF. But bioscientists in the Bay Area did not operate in a vacuum. Without the support of key administrators, like Clark Kerr in the UC system or Frederick Terman at Stanford, the revolutionary changes taking place in the biosciences would have been far less consequential. Only after a program had been established at each Bay Area university did DNA and federal patronage move basic research toward center stage. That it did shaped the overall legacy of the bioscience revolution underway, the organizational framework of each university program, and the particular direction of research itself. Then, as the pace of bioscience research quickened, the issue became no longer whether the biosciences were going to grow at all, but which departments would eventually assume the lead.

By the early 1960s, bioscientists in the Bay Area had good reason to believe that basic bioscience research would continue indefinitely in the mainstream. But behind the autonomy that their expert knowledge afforded them, their phenomenal discoveries, and their seemingly impregnable institutional facade lay serious disciplinary fissures that created an imperfect whole highly susceptible to external shock. As much as Bay Area bioscientists refused to believe it, trouble loomed on the horizon. Their collective decision to take for granted public support—and overemphasize scientific discovery—would have a tremendous effect on the direction and strength of the field in the next generation.

Figure 1. Wendell Stanley, about 1948, just after he won the Nobel Prize for his work with tobacco-mosaic virus. Stanley's mistaken ideas about the virus came from his research on protein. He suggested the latter controlled genes whereas DNA was just a "stupid" molecule. Courtesy of University of California, Berkeley, Bancroft Library.

Figure 2. The Biochemistry and Virus Laboratory, University of California, Berkeley. Courtesy of University of California, Berkeley, Bancroft Library.

Figure 3. Donald Glaser, in the early 1960s. Glaser had a truly remarkable career: Nobel Prize in Physics for invention of the bubble chamber; founder of molecular biology at UC Berkeley; cofounder of the first biotechnology company, Cetus. His career and achievements would continue and include work in psychobiology and the physics of vision. Courtesy of University of California, Berkeley, Bancroft Library.

Figure 4. Joshua Lederberg, around the time of his Nobel Prize, 1958. Never one to focus on a single topic, Lederberg studied a full range of topics, from genetics to exobiology. Courtesy of the Department of Special Collections, Stanford University Libraries.

Figure 5. Stanford University Medical Center in the 1960s. Courtesy of the Department of Special Collections, Stanford University Libraries.

Figure 6. Clark Kerr and John Saunders at the 1962 UC commencement, just before their professional demise a few years later. Courtesy of University of California, Berkeley, Bancroft Library.

Figure 7. UCSF bioscientist C. H. Li, an émigré from communist China. Colleagues wondered if Li's experiments were sloppy, at least until his outcomes fitted with popular and political ideals about applied research. Courtesy of Tom F. Walters and the Kalmanovitz Library, University of California, San Francisco.

Figure 8. Aerial view of the medical towers and hospital at UCSF Medical Center, around the mid-1960s; Mt. Parnassus to the right, Old Kezar Stadium to the left, the San Francisco Bay in the background. Courtesy of the Kalmanovitz Library, University of California, San Francisco.

Figure 9. Biology and Big Science. The biological sciences were once comfortable to people with a background in the physical sciences, but then the field grew large with success and outran basic research questions. The next generation found comfort in more individualized work of genetic engineering. Courtesy of the Ernest Orlando Lawrence Berkeley National Laboratory.

Figure 10. Paul Berg. Stanley Cohen chuckled when he saw Berg complete the first recombinant DNA experiment: "He did it you know, and he—ah, failed— because of, really, the inability to clone it." Berg, nevertheless, won the Nobel Prize for his work with SV-40. Courtesy of the Department of Special Collections, Stanford University Libraries.

Figure 11. William Rutter, who would go on to cofound Chiron, with colleagues, left to right: Raymond Pictet, Axel Ullrich, Rutter, and John Shine. Courtesy of the Kalmanovitz Library, the University of California, San Francisco.

Figure 12. Perhaps without recognizing the patterns of bioscience research, popular culture nevertheless captured systemic changes. Compare just a few political cartoons: the first reflecting cold war–era basic bioscience research, and the second reflecting distrust of the biological and medical establishments in the 1960s. Courtesy of the Washington Post Company and the University of California, San Francisco Synapse.

Figure 13 above and left. "Research life!" and "Put Life First" were ubiquitous slogans among protests in the Bay Area, for they captured perfectly the purpose of those who opposed basic bioscience research. Courtesy of Lenny Siegel and the Pacific Studies Center, Mountain View, California, and the Department of Special Collections, Stanford University Libraries.

Figure 14. A virtual civil war rocked Stanford University throughout April and May 1969 as police in riot gear try to prevent students from entering and disrupting the Stanford University Medical Center. Courtesy of the Department of Special Collections, Stanford University Libraries.

Figure 15. Popular opinion supported the expansion of applied research, but often proved unable to control the forces it had unleashed. These neighborhood children were part of a community-wide effort to halt further expansion of the UCSF Medical Center. Courtesy of the Kalmanovitz Library, University of California, San Francisco.

Figure 16. Vietnam Convocation, May 1970, at UCSF Medical Center. Nixon may have declared "war on cancer," but he was still "the one" whom protesters criticized for not doing enough for medical research. Courtesy of the Kalmanovitz Library, University of California, San Francisco.

Figure 17. A virtuoso of bombast and political opportunity, William Brown (at podium) was a bitter critic of UCSF's propensity to privilege research over medical care. Courtesy of the Kalmanovitz Library, University of California, San Francisco.

Chapter 5
Research Life!

The insularity of the scientific community and its traditional insistence upon sovereignty and subsidy are clearly on the way out. Political necessity now dictates that science must be more responsive to the needs and the tastes of the public.

—Daniel Greenberg

When the 1960s began, at a time when fundamental research questions seemed to dominate the biological sciences at all three Bay Area research universities, a quiet opposition began to question the perceived value of pure knowledge, the federal government's blanket support of research, and the isolation of biology from society's greatest needs. Neither the number of people pressing this cause, the strength of their language, nor their proximate concerns—the general uselessness of fundamental knowledge in the biological sciences—stood out. Their mobilization was small when compared with the vibrant national liberal movements, but they had a significant regional base of strength. Then, by the mid-1960s, isolated discontent erupted into a hailstorm of protest. Pure research, it seemed, was a profound betrayal of the human side of the biological sciences.[1]

Public demand for practical bioscience research was not, of course, the only factor in determining the direction that the field would assume. Nor was a humanitarian ideology ever a uniform, universal, or static ideal. But the broad contours of what could generally be described as "applied bioscience research" remained fairly consistent from about 1959 until 1966, and those ideas played a major role in shaping new federal science policy that would eventually push the field in new directions, and, later on, become the conceptual purpose for a biotechnology industry.

The Cauldron of Pure Research

The sheer scale of postwar federal funding of scientific research through the early 1960s was impressive, its growth downright astonishing. Federal expenditures for research continued to grow at an average annual rate of 20 percent, or to $9 billion annually, although a few elite programs, like those in the Bay Area, fared even better. The budget of the NSF, the primary federal sponsor of basic research, nearly tripled. By 1963, twenty-seven federal programs actively supported scientific research in American universities. The biological sciences had even more reason to celebrate; from 1954 to 1964, the total NIH budgets grew at an astronomical annual rate of almost 25 percent. Between 1950 and 1963, funding of just the NIH's Research Grants Program grew from $6.4 million to $225 million, eventually supporting almost 67,000 senior researchers and more than 35,000 students in training in basic science. Abundantly endowed and protected in its privileged sanctuary where growth could proceed free from public inspection, the biological sciences in universities and medical schools rapidly expanded; new buildings, facilities, and equipment suddenly appeared; and the system churned out a growing supply of graduate and postdoctoral students.[2]

Faith in the importance of pure biological knowledge did not reside just inside the Beltway, or inside Bay Area laboratories. The assumption was present throughout the Bay Area public too. Local newspapers frequently ran special reports updating the public on scientific advances. Local television stations such as KQED invited regional scientists to share with listeners recent research developments. Local editors determined that the crucial issue upon which hinged the outcome of the 1958 10th district Congressional election—Stanford's district—was which candidate would protect federal funding of basic research. One science correspondent, touring Bay Area universities, waxed admiringly that "the enormous, wealthy campuses of Stanford and Berkeley . . . are not only rolling in dollars—they are fizzing with intellectual excitement." The media fed the public's frenzy for details about biological sciences research—"What will they think of next?"—and stressed the amazing ability of these experts to discover a new and powerful future. In the contest between pure and applied research in the biological sciences at all three Bay Area universities during the first two decades following World War II, basic research was the overwhelming victor.[3]

From the preclinical departments at Stanford, from organizational research units such as the Cardiovascular Research Institute at UCSF, and in established research programs at Berkeley, bioscientists in the Bay Area fully expected that the awesome growth of their field would continue, and some even thought their budgets should have been bigger still. In response to one skeptic who wondered if emphasis on the biosciences might in time take funds away from clinical training and weaken patient care, Julius Comroe of UCSF responded dismissively: "not if research experience for the student becomes a *larger* part of the total process of medical education." Speaking at a dinner engagement for the American Foundation, Wendell Stanley called for "*full* support research-wise . . . for the 10 or 20 percent of the brightest men" and responded caustically to doubts about the usefulness of basic research by comparing it to the usefulness of a newborn baby. Russell Lee, a well-respected medical investigator and president of the Stanford Alumni Association, was even more florid in his endorsements, claiming that the "nation is in need, desperate need, of more trained . . . medical researchers." Intoxicated with the apparent contributions of pure science, Lee went a step further: "we need to develop a race of researchers." No one, it seemed, took the time to note that the total federal budget grew during these same years at an annual rate of 6 percent, making it rather easy to see that funding of bioscience research could not continue to expand indefinitely at 25 percent each year.[4]

But in the deep recesses of the field lay troubling professional frustrations. Bioscientists could have reflected on their own discomforting internal situations—how quickly they had expanded their field; how incessant intradepartmental tanglings had disrupted their work; how their unwavering pursuit of fundamental knowledge overlooked practical needs; how the emphasis on scientific research shifted resources away from clinical care; or how they had largely built their empire on the public's money and deference—but they did not. Rather, many in the profession found certain external conditions more troubling: the proliferation and testing of nuclear weapons, the pressures placed upon them by the military to contribute to weapons development, and the vigorous security checks conducted by the relentless House Committee on Un-American Activities (HUAC). Many scientists responded to such unwanted external developments by joining the Federation of American Scientists (FAS), an organization committed to "promoting the welfare of mankind and the achievement of a stable world peace."[5]

It is difficult to pin down the degree of influence that the FAS had in the 1950s, or more significantly, the extent of its influence on protests in the 1960s. Specifically, many tend to believe that the political activism of scientists in the 1950s informed the antiwar protest movement of the 1960s. This argument certainly has merit, but treating the science (physics and chemistry), the issue (nuclear weapons), the strategy (Washington-based political action), and the setting (the cold war) as the template that applies to all scientists exaggerates the broad reach of the FAS. Indeed, the political character of biological scientists at all three Bay Area research universities is marked not by political energy, but by conspicuous political apathy. The decision of local bioscientists to ignore pressing issues in the 1950s—political, social, or otherwise—made them inconsequential as a political body, and more significantly, ill-prepared to contend with or understand the dramatic turn against pure research later in the decade.[6]

The Northern California Association of Scientists: A Political False Start

The most active branch of the national FAS was located in Northern California (NCAS), perhaps because no other region had as many scientists working on or so closely with the Manhattan Project during World War II, where many classified personnel developed a special appreciation for the horrifying power of nuclear weapons. J. Edgar Hoover considered the NCAS "one of the (F.B.I.'s) 25 most important [cases]" and assigned special agents to monitor its meetings. Attendance at NCAS meetings sometimes numbered in the thousands, and local interest spurred the confidence of the organization, radicalizing chapter leaders who directly challenged the developing relationship between the federal government and science in postwar America. The NCAS offspring eventually pushed the ideological boundary of its parent FAS organization, criticizing more forcefully the "proliferation of atomic weapons," advocating "world control of atomic energy," and questioning "government control and direction of research." When *Time* magazine published an article on the FAS, it used NCAS literature to add sensational texture to the piece. Fearing that the "radicalism" of the NCAS might "betray the people," the executive committee of the national FAS often warned Bay Area leaders that they should "not speak too bluntly to journalists." NCAS rarely listened.[7]

But the public listened to the NCAS's aggressive political stances and took it seriously. While it was typical to poke fun at scientists who "did not step out of the ivory tower," they preferred their scientists apolitical and politically silent, and said so quite frequently in local newspapers. NCAS members who advocated peace were considered "menacing," those who pleaded for disarmament were "communists," and those who wanted to share atomic secrets were "traitors." Raymond Lawrence's angry editorial in the *Oakland Tribune* typified the general attack on the NCAS for their radicalism, union-like tactics, and un-American activities:

> Whatever the merits of the controversy between the scientists and the armed services, the [NCAS] scientists have deserted their tradition, method and spirit. . . . The popular notion that the scientific temperament is restrained, judicious, quiet and soundly balanced is belied by some of the recent antics. There is no serene objectivity about this. . . . Here we have a group of extremely angry men speaking. Their passions are as evident as those of John Lewis when he whips up his miners. Dispassionate objectivity was left in the laboratory.[8]

The divide that separated the NCAS and the public created an instant dilemma for biological scientists in the Bay Area. Here was a bona fide organization that had broad professional appeal, but whose radical politics threatened to unsettle the public. Consequently, most Bay Area bioscientists did not embrace what they regarded as "political bluster" and chose, in most cases, to ignore it. Out of the dozens of committee executives and sponsors of the NCAS, only one, Paul Kirk, a biochemist at Berkeley, came from the biological sciences, but even this was a loose designation since he had worked in the Manhattan Project as a chemist. In stark contrast, so many bioscientists in Southern California, Cambridge, and Chicago participated in the FAS that they formed a subcommittee within the parent organization and called it the Federation of American Societies of Experimental Biology—the Northern California branch of the FAS, on the other hand, had no such wing. Sorely disappointed by the lack of bioscientists on the NCAS rolls, representatives from the national FAS made repeated and aggressive attempts to organize "the medical sciences . . . naturalists, physiologists, biological chemists, botanists, biologists, geneticists, and bacteriologists," yet their efforts could be described as nothing less than a dismal failure, as they failed to recruit even one new member.[9]

Their actions were actually worse than ineffective. In the immediate aftermath of *Sputnik*—not coincidentally a peak for cold war hysteria and federal funding of basic research—membership in NCAS dropped from the hundreds to only thirty at Berkeley and nine at Stanford. Eventually, the NCAS at Stanford would fold while Berkeley's survives in name only. Bioscientists in the Bay Area stayed conspicuously silent on other issues too, such as civil rights and social welfare policy, and when pressed, "emphatically oppose[d] 'socialist' forms of health care." On those rare occasions when Bay Area bioscientists participated politically, they did so privately, often in the form of personal letters, much like Howard Schachman in Stanley's BVL, who attached to his signed loyalty oath a private letter that registered his vigorous opposition to HUAC. Sometimes, however, the inaction of local bioscientists resembled more the views of hard-line Cold Warriors than apolitical investigators. For instance, in a thinly veiled allusion, Gunther Stent sarcastically condemned "communist sympathizers" at Berkeley and proclaimed that he "took his daily [loyalty] oath" with pride. Frustrated with Berkeley's sanitized bioscience department, Linus Pauling pleaded with Wendell Stanley to distribute among his staff an antinuclear testing petition, "An Appeal by American Scientists to the Governments and People of the World," an immensely popular document throughout the entire scientific community that constituted a virtual who's who of academia.

Though Pauling and Stanley often corresponded throughout their careers, Stanley evidently ignored Pauling's political request.[10]

There are, perhaps, two ways to think about the inactivity of Bay Area bioscientists in political or social issues. First, this particular group found comfort in the status quo, especially since their field had recently become a major recipient of the federal government's largesse. Furthermore, with little influence on the Manhattan Project's development of the atomic bomb, few had to struggle with the guilt that haunted the physical scientists. Perhaps beneath their newfound confidence brewed extreme anxiousness too, a reasonable concern considering the treatment of politically active scientists: on the eve of Watson and Crick's breakthrough discovery, Linus Pauling was scheduled to go to London to attend a critical meeting on the relationship between DNA and protein, but at the last minute, had his passport pulled by a suspicious State Department. Without specifically singling out bioscientists, Philip Abelson, editor of *Science*, nevertheless captured their noticeable reluctance to engage in controversial issues. "He stirs the enmity of powerful foes," wrote Abelson, and "fears that reprisals may extend beyond him to his institution. Perhaps he fears shadows, but in a day when almost all research institutions are highly dependent on federal funds, prudence seems to dictate silence."[11]

The Pinnacle of Support

After *Sputnik,* bioscientists in the Bay Area never doubted that federal agencies would support their work—a tacit but tactically dangerous endorsement of Congress' ultimate power in science. Two key science-policy advisers expected that government expenditures for university research would continue to increase at similar rates—forever. An NSF executive advised universities to plan as if the agency's budget would grow "at an average rate of about 35 per cent for the next 10 years." Not one, but two executive science policymakers happily envisaged that by the end of the century, the nation might devote half of its gross national product to scientific research. Responding to the suggestion that perhaps the NIH might well do with slightly less funding, James Shannon, an accomplished director of that agency, warned that any cut in his agency's budget "would have disastrous effects on the programs." The insatiable demand for research money took on added meaning when, during Congressional subcommittee hearings, one bioscientist belittled a proposed $100-million increase to his university's budget as "just enough to heat and sweep the various academic laboratory buildings that have been constructed in the last few years."[12]

Biological scientists could make strong claims that they deserved more federal support, especially those in Bay Area laboratories. Indeed, with hindsight it is clear that much of their work had a direct connection to later discoveries, including genetic engineering and cloning techniques. For instance, Heinz Fraenkel-Conrat's work with messenger RNA identified the basic cellular machinery that a later generation would use to synthesize and make copies of DNA. Joshua Lederberg's research with antibody mutations established the possibility that the code for individual genes within DNA sequences were separated by noncoding sequences—a discovery that suggested the technical possibility for future genetic splicing experiments. And Arthur Kornberg's work with the polymerase enzyme provided the technical basis to catalyze DNA replication, expression, and natural recombination.[13]

However, no overall assessment of the state of biological research in the Bay Area during the early 1960s is possible without at least a tentative review of the field's remarkable affluence, and the institutional consequences that it fostered. The worthy purpose of open federal funding certainly provided a sound financial base in which local investigators could conduct fundamental research, but it also allowed for ruinous accommodation of individual autonomy and self-interest. Indeed, some research projects had serious flaws, not only because their original conceptions were shoddy, but also because open patronage created a secure, comfortable environment that protected experimental isolation and independence. Admirers appreciated the Program Project Grant (PPG) through the NIH, and in many ways the PPG was the very model of new federal support for science, for which investigators could double up grants for the same experimental projects—and open dazzling opportunities. Among many examples, W. F. Ganong, an endocrinologist at UCSF, submitted to the NIH a proposal to study "Neural Controlled Endocrine Function," an intentionally broad title that allowed him to apply for more than one federal grant and still have "as much freedom . . . to investigate anything." Ganong's successful PPG received nearly $700,000 per year for thirty consecutive years.[14]

Bioscientists from around the world noted America's proclivity to support research. It was a remarkable and dramatic shift of scientific strength: throughout the first half of the twentieth century, many leading American scientists, such as Wendell Stanley, received advanced training in Europe; but by the 1950s and early 1960s, scientists were considered unqualified until they earned a BTA, or "been-to-America," degree. Among other extreme examples, in 1965, a group of Italian molecular biologists contacted their counterparts at Berkeley and asked if they might be interested in collaborating on a joint research initiative. Perhaps convinced that they had exhausted all opportunities for cooperative research programs on campus—Stanley's BVL never quite took hold—Berkeley bioscientists submitted an "extremely favorable" application to the State Department, the President's Office of Science and Technology, and the NSF to pay for the construction of the "International Stadium of Molecular Biology"—in Naples, Italy.[15]

Of all the affluent experimental programs growing in Bay Area bioscience research laboratories, Joshua Lederberg's understanding of hidden opportunities within the grant system was unmatched. Setting aside his pathbreaking work in molecular genetics, Lederberg applied for and received grant money from every conceivable governmental agency to conduct work in a number of unrelated fields, including biophysics, biohazards, a bioacademic information center, a bio-bookstore, and a program in "renal homotransplantation in man," an impressive piece of nomenclature for what actually constituted the surgical removal of cadaver kidneys and placing them into living human beings. About the time that Kennedy challenged Americans to fly to the moon, Lederberg applied for and received hundreds of thousands of dollars in aid from a number of agencies to begin an "Exobiology" research program—the search for extraterrestrial life forms. Lacking space, Lederberg moved the program into the Kennedy Laboratory for Molecular Medicine, taking valuable space away from clinical research. An exasperated Dean of the Medical School Robert Alway realized he could not reign in Lederberg's "preoccupations" and "personal intervention in extramural matters" if the federal government insisted on funding his "out-of-this-world experimental projects," leading the dean to rhetorically wonder about the "price of Nobel laureates."[16]

An editor for *Science* quietly reflected upon runaway science policy and acknowledged what many preferred to deny: "it is difficult, at best, to form with disinterest and dispassion an opinion about an issue that so vitally affects one's own interests." It was, alas, certainly more than should have been expected from biological scientists in the Bay Area, and elsewhere, as this imperious community seemed at times more concerned with public revenues than public opinion. But their tactics were beginning to attract critical attention, both locally and nationally. The public may have admired the great strides of discovery in recent years, but they also questioned the conspicuous waste—of time, energy, and resources—expended on what many believed were pedantic academic pursuits. As another science editor reluctantly pointed out: "science's ever-growing appetite for money, its unique ways of handling federal funds, and public uncertainty about the payoff it is receiving on its investment in research—all have evoked a good deal of uneasiness." Indeed, by the early 1960s, two powerful forces were moving in opposite directions: the continued growth of basic bioscience research and a popular concern that bioscientists cared little about life in life science.[17]

A Rumble of Discontent

Beyond the realm of science and politics, bioscientists could have heard, if they had cared to, the faint echoes of deepening despair in the domain of public support, that dark region that many investigators took for granted. When Stanford administrators dismissed a popular life-science professor, about a dozen students staged a public protest, asking in collective frustration why the university preferred research rather than teaching and instruction. Bay Area student newspapers may have celebrated the achievements of their two Nobel laureates in the biosciences, but they also ran editorials that noted the lack of adequate health care available to its students: the *Stanford Daily* ran weekly updates of improper medical care administered to the students; UCSF officials actually advised students to avoid the student health clinic because "recent injuries there have indicated that . . . an emergency visit to the Student Health Facility is potentially dangerous." Concerned about nuclear proliferation, a handful of students on all three Bay Area university campuses joined the Committee for a Sane Nuclear Policy (SANE), and Stanford and Berkeley students flocked to hear the comedian Tom Lehrer poke fun at the proliferation of weapons of mass destruction ("So long, Mom, I'm off to drop the bomb, So don't wait up for me"). On occasion, a lonely voice would rail forcefully about the collective priorities of the nation toward health care, such as Jim Lieberman, student body president at UCSF, who offered a general yet poignant attack that linked the medical and military establishments:

> It is difficult to live by . . . the Hippocratic Oath: to watch a cancer patient waste away, to nurture imbeciles and idiots and speechless spastic children; to tube-feed senseless human vegetables who lie in bed for months. . . . It is difficult, but a fitting part of a doctor's reverence for life, and his integrity depends on that. But integrity demands much more than that from him and every human being. . . . I must not take a life no matter how I feel about it. . . . But look what follows! War and capital punishment must go, and armies must disband. For I insist that I am morally responsible for pulling the trigger of a loaded weapon, even when it is aimed at someone designated as my enemy. If I cannot judge the lives of those I know best, how dare I take the life of someone new, someone I'm supposed to meet in battle because our governments have failed to keep the peace.[18]

Broadly speaking, the public could legitimately claim that its impatience with science descended from historic criticisms that originated after World War II, when postwar analyses offered idealized reappraisals of atomic power, nuclear proliferation, and doomsday predictions by scientific organizations such as FAS. By the end of the 1950s, these modest criticisms of science showed signs of wriggling free from their traditional restraints, and that a nascent anti–basic-science subculture would eventually become central to the political culture of the next generation. But these early criticisms typically missed their mark, primarily because they lacked focus and could not penetrate scientists' trademark mix of pride, prejudice, and productivity.

Public criticism of science probably would have remained isolated in its effectiveness, and by itself probably would have failed to effect major changes in the biosciences, at least in the short run. But basic bioscience research was not an ordinary issue for those harboring doubts. Throughout the Bay Area, both among those who were beginning to question scientific autonomy and those who believed in its absolute value, passions were mounting that would wreck laboratory isolation, antagonize interdisciplinary tensions, and confront individual investigators with a popular crisis they wanted so desperately to avoid. The two sides would test the very fabric of the biological sciences—and their combined efforts would eventually push the entire field in new experimental directions.

Public opinion began to turn in 1959, when local and national newspapers ran front-page stories about a legal morning-sickness drug that caused horrible physical birth defects. Readers learned about a woman by the name of Mrs. Finkbine, whose doctor prescribed thalidomide as a sedative that would "tame the rough edges of her pregnancy." But complications immediately arose, and Mrs. Finkbine's doctor had no answers. Concerned about her worsening condition, Mrs. Finkbine traveled to Sweden to get an abortion, only to find that she was carrying a severely deformed fetus. As word about Mrs. Finkbine spread throughout the medical community, other horror stories that linked thalidomide to birth defects began to emerge. Newspapers responded to overwhelming interest in the story with a series of follow-up articles about Mrs. Finkbine's nightmare and the lack of certified testing of the drug by any qualified physician; local papers like the Hearst-owned *San Francisco Examiner* described in shocking detail the phocomelic deformities that the drug could cause in children. The sheer magnitude of the disaster also caused widespread panic: Merrell Company distributed thalidomide to more than 1,200 doctors across America; nearly 16,000 women received prescriptions, of which 624 were in the first trimester of pregnancy. Worse, the medical community found it nearly impossible to recall the drug once it had been dispensed. Perceptive readers feared the worst: that a significant number of women had already taken the dangerous drug.[19]

The thalidomide scare—what one scholar describes as "the twentieth century's worst drug disaster"—had a powerful effect on the Bay Area psyche, which was becoming especially sensitive to this and other scientific mishaps. Alongside thalidomide stories, muckraking journalists ran articles about Cutter Laboratories in the San Francisco East Bay, which had been accused of distributing a polio vaccine that induced rather than prevented the disease. The contrast in terms of how the public interpreted the polio outbreak and how local bioscientists defended Cutter Labs' negligence could not have been more striking. On the one hand, the public reacted with horror to the before-and-after pictures of once healthy children who had innocently taken Cutter Labs' polio vaccine. On the other hand, according to lab founder Bill Cutter, the expert testimony of two Berkeley bioscientists proved critical to the defense of the company, allowing them to settle the damages out of court and escape certain bankruptcy. A handful of editorials that appeared in local newspapers described the expert witnesses from Berkeley as "irresponsible scientists," while most agreed that the testimony strained to the breaking point the local community's tolerance for unresponsive science.[20]

Blows to the familiar continued. In the early 1960s, the Surgeon General of the United States reversed course and issued the first formal statement that cigarette smoking had harmful medical consequences. At the same time, a group of local chemists publicly acknowledged that the South Bay's fluoridated water experiments could inadvertently damage some children's digestive tracts. Newspapers in the Bay Area and around the country also published photos of animals poisoned by pesticides used against "invading" fire ants. The public barely had time to process such news when a report came out that Frances Kelsey, a drug reviewer in the FDA, fought single-handedly against her superiors and risked losing her career in order to remove thalidomide from the American market—even after a relationship between the drug and Mrs. Finkbine's deformed fetus had been established. Expectant mothers contacted their doctors to find out if there were any other drugs that might inadvertently harm their babies, and neighbors spoke as if hundreds of children in their own communities might have polio.[21]

Public outrage gained momentum when Rachel Carson's *Silent Spring* appeared in 1962, raising serious questions about the objectivity of expert scientists. Carson's hard look at the effects of insecticides and pesticides on songbird populations throughout the United States, whose declining numbers yielded the silence to which her title referred, set off a wave of public disgust and galvanized a nascent ecological movement. Other ecological critiques of science eventually appeared, such as Charles Reich's *The Greening of America*, which became a staple of the environmentally concerned protest culture, and Lewis Mumford's *The Myth of the Machine*, which assailed technological achievements as harmful to human life and values. Along with these poignant essays sprang other general-science studies, such as Thomas Kuhn's *The Structure of Sci-*

entific Revolutions, which questioned "internal" studies of scientific discoveries that did not also take into account historical context. Then came a series of sociological studies of scientific knowledge, such as Robert Young's essays on biological and social theory, and renewed interest in the exploration of the relationship of science and society in such works as Boris Hessen's *Science at the Cross Roads* and J. D. Bernal's *The Social Function of Science.* At the time, critique of science was considered radical and activist in origin; other than Carson's and Kuhn's work, most found small audiences in such notoriously radical literary courses taught by the poet-scholar Thomas Parkinson at Berkeley, or H. Bruce Franklin at Stanford, or among the bohemians in UCSF's neighboring—and notorious—Haight-Ashbury district. Of these science studies, it was Carson's book that enjoyed the most success in the Bay Area, because her discussion of toxic pollutants challenged the region to push its environmental values beyond local concerns and protect the health and well-being of the environment. Quality of life as an ideal and as a focus of public action lay at the heart of what was just beginning to emerge in the Bay Area.[22]

Then came Vietnam. So much has been written about the war and the protest against it that it would be redundant to cover it here; suffice it to say that popular memory overshadows the subtle humanistic questions that the war raised among many. Beneath the historical skepticism, suspicion, and confrontation that many people commonly use to describe this era, the war and its debates had important consequences for ideological humanists, especially in the Bay Area. American military commitments in Southeast Asia certainly unsettled a large number of people, many of whom had become in recent years openly skeptical about the certainty of human life. Escalation catalyzed the Bay Area, too, and created a concentration of protest that was especially powerful in San Francisco, Palo Alto, and Berkeley, where activists initially reacted with modest "teach-ins," peace rallies, and petitions. Although the spectrum of concerns about America's involvement in Vietnam spanned a wide variety of issues, a central debate at the time was a humanist one: whether a technologically advanced society using sophisticated weaponry, such as napalm or other kinds of biochemical warfare, could or should fight war in an impoverished region.[23]

Bay Area bioscientists, who had only recently enjoyed near total deference from the local community, responded to the storm of fury now spinning well beyond what they could control with defensive and ill-conceived remarks. Reacting to the rising tide of protest, Manuel Morales, an investigator in UCSF's CVRI, boasted that "any system with an intellectual basis will outlast one based on 'democracy.'" Stanford's Joshua Lederberg warned an audience that the greatest threat to American health was a government that took into account public opinion: "a well-intentioned government might impose rash commitments for the sake of short-term advantages." Struggling to understand why medical schools should attend to patients rather than conduct fundamental research, Arthur Kornberg reflected: "the engineering department doesn't do engineering; the law department doesn't do legal work." Heinz Fraenkel-Conrat in Berkeley's molecular biology department dismissed popular concerns as simply following the "fashionable party line" and recommended that anyone deeply frustrated might be better served if they took "any private problems they might find of sufficient magnitude to their spiritual advisors or psychiatrists."[24]

A proverb warns of pride preceding a fall. One simple comparison illustrates the rapidly widening gulf between the public and the profession. In early 1959, just prior to the thalidomide scare, most Americans looked upon their scientists with unwavering admiration; bioscientists in particular seemed nearly omnipotent. A Gallup poll conducted at the time confirms an illustrious image: a majority of Americans considered scientists, doctors, and medical researchers the second most prestigious profession, behind Supreme Court Justices. Medical researchers had followed their successes with penicillin and streptomycin in the 1940s with a plethora of new drugs, including antihistamines, cortisone, and more antibiotics in the 1950s; by 1959, 80 percent of the drugs being prescribed had reached the market in the previous fifteen years. Vaccines greatly reduced the incidence and mortality of whooping cough and diphtheria, and the public anticipated similar successes in the effort to prevent or control mumps, measles, and rubella. And newspapers reported almost daily the remarkable fundamental discoveries seemingly pouring out of bioscience laboratories. Nothing, however, did more to enhance the status of medical research than the fight against poliomyelitis. Polio struck children and young people at random, sometimes killing them, but more often simply leaving victims paralyzed or trapped and barely alive in "iron lungs." A crash research program sponsored by the federal government paid off handsomely when Jonas Salk of the University of Pittsburgh Medical School developed a virus vaccine against the disease, and then again with government help, mounted a nationwide inoculation campaign in 1954. In spring 1959, polio was no longer a major concern.[25]

But five years later, in late 1964, the thalidomide scare, the Cutter Labs' polio outbreak, *Silent Spring*, fluoridation, insecticides, Vietnam, and other related concerns contributed to the public's growing dissatisfaction with science. In a sense, the field had become a victim of its own successes. Happily taking credit for all of the amazing advances in their field and speaking confidently to the public about their planned offensive against heart disease and cancer, bioscientists had actually made little headway in the fight against the primary killers of the American middle-aged and middle class. The public proclamations by these biomedical experts—which can be dismissed as simply innocent rushes of enthusiasm—nevertheless established expectations impossible to fulfill, which further eroded their prestige in the eyes of an impatient public. Making matters worse, too many researchers compromised themselves and their profession by continuing to tout the safety of cigarettes and radiation in ads and public forums. Adding to the frustration, recent scientific breakthroughs led to rising medical and health insurance costs for millions of Americans who could not afford many of the miracle treatments that physicians claimed were now available. Pollsters found public confidence in biological and medical research rapidly falling in 1964, ranking medical researchers far lower in importance than they had in 1959, and bottoming out with a 37 percent approval rating soon after. The murmur of discontent that bioscientists had chosen to ignore was now threatening to swell into an insistent cry for immediate change.[26]

Bioscientists had woefully underestimated rising popular frustrations and had profoundly miscalculated their decision to dismiss early public criticisms, especially since it was the taxpayer who ultimately paid for their "molecular empires." But the irony of their pride runs deeper, for it was in their very own backyard where they would have found the strongest contempt for their work. By the mid-1960s, young people enrolled in or living near the three Bay Area universities were becoming politically powerful, volatile, and restless opponents, and their all-out assault on the establishment put enormous pressure on all institutions, including the biosciences. The passionate dissatisfaction that resonated on each campus catalyzed an eclectic mix of political allies. By no means did all three campuses speak in unison or experience equal degrees of unrest in the mid-1960s; however, just beneath the scattered concerns about the impracticality of fundamental bioscience research ran a highly sensitized desire to humanize it. For many students, no field held greater promise than the life sciences.

Student Protest and the Emerging Political Culture

When the radical twenty-one-year-old graduate student Mario Savio climbed the steps of Sproul Hall during Berkeley's eventful fall semester 1964 and railed against the establishment, it was not at all obvious at the time that the steady, escalating protest about which he spoke would also exert pressure on the biological sciences:

> We have an autocracy which runs the university system . . . and the faculty are a bunch of employees and we're (students) the raw materials . . . who don't mean to be made into any product, don't mean to end up being bought by some client of the university, be they the government, be they industry, . . . be they anyone. We're human beings. There is a time when the operation of the machine becomes so odious, makes you so sick at heart that . . . you can't even passively take part, and you've got to . . . indicate to the people who run it and to the people who own it, that unless you're free, the machine will be prevented from working at all.[27]

Many observers believed, then and now, that Savio was simply speaking to and for a popular front of leftists, bohemians, and folksingers, and that his general protest was fixated on the advancement of libertarian issues like the free speech movement. Perhaps, but behind the wide spectrum of political and social issues voiced by activists such as Savio, it is possible to detect early signs of a restive humanitarian strain of concerns. Indeed, if anything united the disparity of protest in the first half of the decade, it was a desire to achieve a moral society and instill greater respect for life. During an all-night vigil in which hundreds of local university students stood outside a prison in San Rafael to protest capital punishment, one demonstrator sadly reflected the sentiment he shared with his peers: "how does [this] work? . . . I mean, no one wants to see [prisoners] die." Even the lonely student who sat quietly on Telegraph Avenue just off the Berkeley campus could make a powerful humanitarian point simply by wearing a gas mask and a sign that read: "buy your gas mask now—prepare for chemical warfare in World War III." Student slogans such as "Not With My Life You Don't," "War on Poverty—Not on People," and "Make Hell, Not War" tapped a deep vein of concern, what one student described as a battle against "man's capacity for inhumanity." Speaking for many, editors for the *Daily Californian* determined that "man's survival . . . depends upon immediate humanitarian uplift."[28]

Even though few protesters could explain precisely what "humanitarian uplift" might look like, the hell that they were making in the streets was nevertheless making inroads, especially among Berkeley's student body. Perhaps most telling of their effectiveness was not necessarily the ferocity of the protesters' rhetoric but the sentiment of students observing the action. Between 1963 and 1967, virtually all bioscience disciplines at Berkeley experienced flat or declining enrollments—sometimes by as much as a 24 percent drop—while student enrollments in all other academic departments grew—including the physical sciences—generally by an average of 3 to 5 percent each year. Sensing the notable shift but struggling to understand it, one student astutely wondered if his peers in the physical sciences were simply more comfortable that the "theoretical had driven out the applied" than those who might have interest in the life sciences. One Berkeley broadside encouraged students in the biological sciences to "find issues on the local level, in the community, around which people can demand control." For many, bioscience research had become that issue because its fluid experimental frontier—suspended precariously between impractical "pure" scientific fields such as physics and chemistry on the one side, and "utilitarian" medical or agricultural research on the other—made the discipline capable of either contributing to human misery or delivering ultimate salvation.[29]

In Berkeley's heated atmosphere of escalating political and cultural ferment, the broad contours of not one, but two strains of concern about the biological sciences began to take shape. One condemned all science and scientific issues. These notorious antiscience groups, such as the Diggers, actually practiced a kind of philosophical anarchism rooted in individualism and opposed virtually any American institution, including but certainly not limited to science. More focused than the Diggers were other radical antiscience groups who specifically condemned all science, no matter its objective, as imperialistic. Earth Day attacked science as the primary tool of exploitation for the class in power and insisted, in essence, that science and a healthy environment could not coexist. These Earth Day advocates used the Vietnam War as their key piece of evidence to "prove" their apocalyptic vision: "science has fueled the war and . . . the war has destroyed the environment." Perhaps the most dedicated opponent of the biological sciences was Science for Public Death, which attributed existing evil to "the gradual perversion of advances in those biological sciences to which it has historically entrusted the promotion of human life—medicine, agronomy, pharmacology and biochemistry." In short, the group believed that the biological sciences promoted public death—hence their name—and pointed an accusing finger at bioscientists "who had contributed to the degradation of the life sciences." To them, the development of biological weapons demonstrated "how readily basic research could be subverted," and they made a point of "damning" companies like Cutter Labs in the East Bay for carelessly producing "plague vaccines." Although these antiscience groups yearned and fought, sometimes courageously, for an easier and simpler life, their protest remained small and secluded, condemned to fight against a university system whose commitments to scientific research ran deep.[30]

Yet the isolation of antiscience protest still left plenty of room for radicalism—a peculiar style of radicalism that accepted the benefits of scientific research at face value but wanted to redirect it toward humanitarian objectives. It was a powerful channel of ferment that could attract on occasion the sympathies of moderates, but whether it could establish and then sustain that momentum remained to be seen. The most potent of the utilitarian science groups was the Society for Social Responsibility in Science, Scientific Workers for Social Action, Environmental Action for Survival, and the most influential, Science for the People and its bizarre list of peripheral organizations. Their techniques may have seemed like daft attempts to promote public awareness about the misuse of bioscience research, but rarely did they offend public sensibilities. For instance, a splinter group of Science for the People, the California Institute of Man in Nature, designed "beautiful buses—a new form of schoolroom" and covered them with drawings of "rivers, birds, trees, grasses and reeds, and flying pennants proclaiming the earth's fertility." Friends of the Earth offered free samples of .07-milligram tablets of DDT—the maximum daily allowance as mandated by federal law—as a shocking display of unchecked science. Ecology Action had perhaps the most sophisticated understanding of the biological sciences when it condemned the NIH for its lack of responsiveness to public health and demanded that the organization support "science and technology that can be used to end hunger, population explosions and other evil." Their massive picket line that strung out across four city blocks in downtown San Francisco suggests the potential strength of this group.[31]

However, beneath the groundswell of public support for more responsive bioscience research lay a precarious union of interests that was capable of unraveling at any moment. A raucous debate between Berkeley graduate student Mark Schechner, the culture critic and visiting scholar from Columbia Jacques Barzun, and the neoconservative Harvard sociologist and visiting professor Nathan Glazer captures in the extreme the uncertain future of bioscience activism.

Their three-way debate began in late November 1965 when Mark Schechner wrote a provocative article for the Berkeley student newspaper in which he described the university's "ivory tower . . . looking more like a tree growing in a cemetery—tall perhaps and green but rooted, after all, in death." Energized by what he described as a utopian "vision of better things and a sense of personal responsibility for the world," Schechner called for total and immediate divesture by the University of California from all research sponsored by the federal government. Days later, Jacques Barzun responded with an article that waxed eloquently about Schechner's insightful treatment of the "inalienable" relationship between science and society. Barzun made it clear that not *all* federally funded scientific research projects should be condemned—only those that promoted fundamental research: "we (the public) have used and loved science—when I say science I mean 'pure' science—unwisely." A testament to the enormous interest in the debate, Nathan

Glazer's neoconservative response appeared after winter break, but the delay did not soften his criticism of Schechner, described as a naive and irresponsible desire to "divest" from all federal contracts, or of Barzun's "painful and tiresome attempt" to reject all fundamental research. Instead, Glazer proposed that each fundamental research project should be evaluated according to how it might "deal with the practical questions of society." Finally, the editor of the *Daily Californian* offered a reflective piece, which celebrated the "common desire" shared by all three participants to redirect scientific research toward more humane ends, an idealistic assessment which in reality obscured a central truth: behind the apparent "common interest" shared between the three commentators lay a spectrum of differences: opposition to all federal patronage, all pure research, or just identifiably irrelevant pure research. How these opinions played against each other on the Bay Area's larger stage would eventually shape future protest—and, ultimately, the direction of the biological sciences in the coming years.[32]

Stanford activists may have responded to questions about the biological sciences once the issue had become a significant concern at Berkeley, but the delay in no way defines the Palo Alto campus as a bastion of conservatism. In the mid-1960s, a growing number of New Leftists at Stanford turned against the traditions of wealthy alumni, stood their ground against the authority of their administration, and even earned a degree of respect and sympathy from many of their suspicious classmates. Because the majority of students at Stanford came from white, middle-class families, many of whom categorically rejected any radical who disrupted class or condemned a war that their nation was fighting, activists on campus developed what must have seemed like multiple personalities: they clung to more traditional ideals of peace, talked in the open like Berkeley liberals, and saved their bolder, confrontational rhetoric for more private gatherings. In general, Stanford students practiced a more voyeuristic form of activism than their Berkeley counterparts. For instance, while hundreds of Berkeley students stood in defiance with other screaming, singing demonstrators against HUAC interrogations at a San Francisco City Hall trial in spring 1960, fifteen Stanford students "stayed silent," preferring to observe the demonstrations from a distance. Days later at a showing of HUAC's controversial film "Operation Abolition" in Cubberley Auditorium on the Stanford campus, a raucous, overflow crowd criticized HUAC for its unconstitutional abuse of civil liberties; however, the protest fell on deaf ears since no one from HUAC was actually in attendance at the time.[33]

Given the early docility of the Stanford student body, university administrators and its science faculty summarily ignored the isolated disruptions and maintained their commitment to federal funding of science and strict devotion to basic research. In 1960, Stanford agreed to allow the Atomic Energy Commission (AEC) to build a massive, $153-million linear accelerator (SLAC) in the rolling Stanford foothills directly behind the campus. Months later, the Defense Department's Advanced Research Project Agency awarded Stanford a $10-million contract to help overcome the school's "lag in basic research materials." At about the same time, the university procured millions of dollars of support from various federal agencies and private foundations to upgrade its biophysics laboratory by "synthesizing [it] with physics, physical chemistry, and electronics," a decision and a direction that spoke volumes about how Stanford officials viewed at that time the proper direction and purpose of bioscience research.[34]

What Stanford officials failed to recognize was that the community's passivity had already begun to recede. More and more people were beginning to wonder about federal policy that supported scientific research, especially on at least one front: the efficacy of uninterrupted growth—of nuclear weapon development, of military science research, and its destructive applications. Stanford physicists defended President John F. Kennedy's decision to increase the defense budget and resume nuclear testing following the Soviets' detonation of a massive nuclear bomb in 1961, and then faced unrelenting criticism from the student body for their views. Students also turned on none other than the university's de facto visionary Frederick Terman when he insisted that "the US must increase its number of highly trained engineers and scientists in order to compete with the Russians." And a jittery Stanford audience doubted the opinions of AEC physicists who claimed that there was "absolutely no radiation threat" from SLAC; the audience may have lacked data to support their opposition, but they certainly did not lack evidence that Stanford was generally more committed to fundamental research than it had been in the recent past. Indeed, University President Wallace Sterling convinced no one when he responded to growing criticism by arguing that the massive growth of federally funded fundamental research at Stanford was good for everyone:

In 1940, the dollar figure (for federal support of research) was 74,000,000.

In 1950, the year in which the Korean War began, the sum had increased almost 15-fold to 1,083,000,000. And in 1964, more than 200-fold to almost 15 billion. . . . That the availability of these sums has had an impact on our . . . university there can be no doubt. These research funds have brought to distinguished university professors a number of things: opportunity to pursue an important intellectual interest; status that is associated with research grants; brilliant graduate students who are attracted to the professor both by his distinction and by the research grant at his disposal; and consulting opportunities which, according to one observation, may "add variety to his life and dollars to his income."

Clearly, Sterling's message signaled that the university had no intention of recognizing the growing concerns, easing its relationship with the federal government, or endorsing a change in priorities that would include more practical research projects.[35]

Students struck back at Stanford's callous dismissal of their concerns, and some turned their attention toward the biological sciences as a field in need of change. Much like the enrollment trends at Berkeley during the mid-1960s, undergraduate and graduate students at Stanford avoided the life sciences, again a conspicuous development because it stands in marked contrast with the enrollment increases in all other disciplines, including the physical sciences. In activities reminiscent of humanistic protests waged across the bay, students at Stanford publicly condemned the Army's plans to emphasize chemical and biological warfare research, turned out in the hundreds to protest nuclear weapon tests, circulated fliers that had a decidedly moral tone such as "Clean Milk and Dirty Bombs Don't Mix" and "Thou Shalt Not Kill," and even petitioned the university to stop using toxic chemicals in the campus roach control program. Not to be outdone, the other side of the political spectrum got involved too, as Young Republicans, University Society of Individualists, and Birchers collaborated on a pamphlet titled "Man and State," which called for, among other things, massive and immediate reduction of government-subsidized research. Most of the activists who considered the biological sciences a legitimate target could hardly be described as radical—they were, more accurately, more restrained than their notorious brethren—though a passionate few were willing to take the issue to its extreme. For instance, an anonymous group of activists cut SLAC's electrical power and telephone lines and burned the facility's water tower in order to "stop the production of radioactive material [by SLAC] and potential damage to the local ecosystem."[36]

Perhaps the first direct attempt to humanize the biological sciences at Stanford came in the form of a two-pronged attack waged by medical students against the university's highly touted five-year medical school program. Activists sent a petition to the California Academy of General Practitioners, claiming that the production of "gold-plated super-atomic medical specialists . . . excluded the family doctor," and staged a passionate demonstration outside the medical center to protest the dehumanizing tendencies of any program that emphasized research over patient care. Although their activities had little immediate effect—the State Association of General Practitioners offered obligatory support and urged university officials to reconsider the benefits of the traditional four-year curriculum—university medical officials cautiously directed a panel of local medical professionals to take student concerns seriously and proposed a new medical training program that reduced the research requirements of the five-year program.[37]

Yet the more remarkable thing, as striking as Stanford medical students' first direct action against the complacent princelings in the biological sciences, is how quickly fratricidal war could break out between radical activists and moderate sympathizers. On 31 October 1965, a group of medical students split off from the university's largest protest organization, the Stanford Committee for Peace in Vietnam (SCPV), and formed the Medical Aid Committee for Vietnam (MACV). The objective of the MACV was to use more radical platforms in which to "protest the particular and probably un-payable debt to the civilians whom the war has harmed." The organization of the MACV on the Stanford campus was astonishing enough. More astonishing, the MACV held a blood drive in White Plaza to help civilian casualties of U.S. bombing in North Vietnam and the National Liberation Front (Vietcong) areas of South Vietnam. Many contemplated the MACV's blood drive with a distaste that bordered on horror: as moderate sympathizers saw it, there was something mean and violent about a movement that wanted to protect the health of the Vietcong. Student leaders in the more moderate SCPV quickly organized their own protest alongside their radical MACV brethren with a massive and more centrist protest against the use of napalm.[38]

The combined effect of the two protests was staggering, and local newspapers swarmed the Stanford campus to cover the story. Hoping to cushion the blow, executives at United Technology Center, a local manufacturer of napalm, sent a spokesperson to the Stanford campus to defend the company's actions. "Everyone was responsible" for the horrors of war, he argued, not just United Technology, and the company's "ultimate goal [was] to halt the production and use of napalm altogether." His attempt to assuage protesters backfired. Instead, activists and the community rallied against the company's duplicitous defense and days later staged the largest demonstration in Stanford history. Symbolic of how much had changed since the days when Stanford bioscientists refused to join the Federation of American Scientists in the late 1950s, eight bioscientists at Stanford participated in the march and seven signed the SCPV's anti-napalm petition.[39]

Protests against napalm helped transform a divided campus into a center of political activism. A few months after the demonstration and just six months after one poll found that Stanford students supported using military force in Vietnam, a new poll found that a majority of students now favored immediate "de-escalation." Moreover, after almost twenty years of guiding Stanford University into a lucrative relationship with the federal government, and after helping shape Stanford as one of the elite research universities in the country, Frederick Terman decided to retire from his position as vice president and provost, yet had enough time and energy to continue as a director of Hewlett Packard, Ampex, and Watkins-Johnson Corporation, act as vice president of Stanford Research Institute, and serve as a consultant for numerous governmental and educational agencies. While most of the Stanford community was deeply saddened by the retirement of this great administrator, a number of protest leaders quietly gloated that they had driven out "the most dangerous man at Stanford."[40]

Of all the attempts by activists to force the biological sciences to become more relevant during the 1960s, the most confounding were the efforts put forth by students at UCSF. The location of this urban medical center—its neighbors were the infamous hippies of the Haight-Ashbury district—and the student demographics—all graduate students enrolled in professional or preprofessional medical training programs—created a unique contrast in which an older, conservative student body committed to a specific and rigorous vocational path confronted on a daily basis perhaps the most intense counterculture in the country. "Students at the [UCSF] Medical Center," observed one perceptive student, were "far less tolerant of aberrant political views and activities and also of more general social non-conformity" than those in the "somewhat 'ivory tower' academic world." "Medical students [at UCSF]," remarked another, have a dominant "conservative character," where "apathy" has become the single-most important "dilemma faced by the health science student." Indeed, the peculiar withdrawal by UCSF students from broader social issues was made more peculiar by the fact that many entered the profession because, as one student put it, "most were sincerely concerned about people's welfare."[41]

As late as 1963, UCSF's participation in the broader protest movement could at best be described as reserved, especially when compared with student activism at Stanford and radicalism at Berkeley. While Stanford students staged massive sit-ins to block the entrance of the university's administration buildings, UCSF students conducted surveys on cafeteria food and found that most "preferred shortcake for dessert." While Berkeley students took turns shouting "fuck" into a megaphone while standing on top of a police-car-turned-soapbox to defend the "Filthy Speech Movement," UCSF students accepted the administration's decree that only faculty and staff could use the main cafeteria. When a majority of students at Berkeley and Stanford believed that America should withdraw, on some level, from Vietnam, a majority of those polled at UCSF still favored U.S. involvement and even supported, in special cases, the bombing of civilian areas.[42]

Certainly, events that took place off campus gradually forced a reluctant student body to reconsider their lifeless engagement in world affairs, but none had as dramatic an effect on the student body as a whole than two unintentionally provocative articles in the student newspaper. Roger Lang, editor of the student newspaper *Synapse*, used President Kennedy's assassination to write a highly controversial article in which he pointed an accusatory finger at "extremists of all types" who, he believed, "might do well to pause and wonder if their inflammatory speeches and documents might . . . push fanatic[s] over the brink to murder." The UCSF student body, many of whom were anything but radical, disagreed vehemently with the characterization put forth by the student editor and rejected any notion that there was a link between civil disobedience and Kennedy's assassination. The furor over Lang's ill-conceived article became so intense that he resigned as editor. Soon after, the next editor of the student newspaper, Melvyn Matsushima, offered an equally grave opinion that the whole of the student body at UCSF had "confidence in President Kerr and strongly endorsed the maintenance of law and order on University of California [Berkeley] campus." Much like Lang, Matsushima also misinterpreted UCSF's passivity for compliance. And while Lang's article may have awoken a sleeping student body, it was Matsushima's ill-timed article that galvanized an angry student body and energized the heretofore inactive protest culture. Matsushima, overwhelmed by the response against his article, also resigned and was replaced by a new and more sympathetic editor who wisely suggested that perhaps readers should "explore the ideas of the boat-rockers, and weigh them carefully."[43]

The two controversial student articles sparked a campuswide debate and helped push UCSF students far beyond the issues raised by Lang and Matsushima. Much like early debates at Berkeley and Stanford, the range of immediate concerns at UCSF had no principal focus. But by 1965, most conferences, speakers, debates, and forums began to address a single question: the difference between the *art* and *science* of medicine.

Most of the student body at UCSF believed strongly that the "art of patient care" was the touchstone of medicine, and that the analytical side, symbolized by fundamental research, was well established and overemphasized. In general, most students agreed that current conditions dictated an immediate need to elevate "the interest of the patient as a whole personality, not as a disease entity." Books such as *On Becoming a Person* by Carl Rogers, which advocated patient-centered medical care, and Richard Carter's *The Doctor Business*, which urged patients to spend their health-care dollars wisely, became instantly popular on campus.

Students, more conscious of their professional relationship with each other, challenged professors who graded on a curve because "competition made the entire medical profession more individualistic," because such a method of evaluation "ruined morale," and because "a single 'brain' is not really as worthwhile [to society] as having a good class in general." More important, UCSF students branched out and became more engaged in off-campus activities. For instance, dozens of UCSF medical students joined the Student Health Organization chapter on campus and offered such services as the Skid Row Medical Clinic on Fourth Street in downtown San Francisco; others formed the Committee on Problems of War and Peace and began collaborating with their more radical counterparts at Stanford and Berkeley.[44]

In general, by 1965, many UCSF students had determined that the university had overindulged in research, the application of science to treat disease, and the search for and dependence upon federal patronage. The combined effect of student concerns produced a powerful force that wanted to rearrange the doctor-patient relationship and trim the inherent dangers associated with a medical community that ignored the sick. For instance, health activists at UCSF shared with social welfare advocates' similar worries about the rising costs of medical care and the poor service that sick patients typically received. But these patient-centered activists also believed that the perverted priorities of biological scientists had infected their vaunted medical profession, where too many physicians focused on research rather than treatment and prevention, had little time or interest to improve their communication skills with patients, and developed all-too-close ties with money interests like insurance and pharmaceutical companies or the AMA. Patient-centered protest on the UCSF campus, once weakened by the inherent conservatism of an older student body, had laid the groundwork for a deeply suspicious student body that now sought an issue in which they could exact real change.

That issue arrived in May 1965, when the medical center admitted a Bay Area woman who had traveled to Tijuana, Mexico, for a "south-of-the-border holiday"—the code phrase for an illegal abortion. Apparently, inside the car-turned-hospital, the woman and her husband paid $800 to have her fetus removed. By all accounts, it was a gruesome procedure that entailed all the accoutrements of what should be expected of any illegal medical operation: the "surgeon" had no formal medical training, no surgical instruments, and did not use anesthesia. Nearly three months of hemorrhaging later, she came to the UCSF Medical Center to have corrective surgery. One of approximately 80,000 similar stories each year, this woman was especially lucky; in 1964, nearly 1,000 women died of complications that resulted from an illegal abortion.

No one at UCSF denied the misfortune of abortion. Nor did anyone deny the proposition that if a woman insisted upon an abortion, she would be much better served having it done by a trained professional in an appropriate setting. Before spring 1965, there was something quaintly anachronistic about the physicians' traditional faith in the state's 1873 abortion law and their attempts to treat women who had been mutilated by an untrained "doctor" without asking any questions. But the relative importance of creating a more humane medical center had grown, catalyzing around a single issue: preventing whatever had caused this woman to nearly die. The collective response by the UCSF community to the situation was staggering, and it was a classic case of massive resistance. Between January and May 1966, UCSF physicians reportedly performed fifty-six illegal abortions, forty-six in response to an illness that the expectant mother had contracted—typically rubella or chicken pox—and yet was technically illegal under existing legislation; however, an informal poll estimated that the number of abortions performed was much higher, while the number of pregnant women who were sick was much lower. All of the operations were screened by the hospital's therapeutic abortion committees, senior staff of the OB-GYN department approved each procedure, the resident staff were aware of the program, consultants from other disciplines provided advice, medical students usually observed at least one of the procedures during their rounds of training, and most of the rest of the student body had a strong suspicion that abortions were performed on the premises. One local newspaper proclaimed UCSF the nation's "abortion capital."

Five months into the program, the State Board of Medical Examiners brought charges prepared by the Attorney General's office against two UCSF obstetricians for violating the 1873 abortion law. In a dramatic show of support for their colleagues, seven other physicians confessed, and they too were charged. Shortly after the charges were presented, the San Francisco Medical Society, made up primarily of physicians from UCSF, voted 94 to 9 "in support of the actions of these physicians," and attached to their statement a petition that contained 132 signatures of medical students who informed the state Attorney General that they too defended the actions of the accused and would, in the future, perform abortions. So overwhelming was opposition to the state's abortion law that during pretrial hearings the judge spared little time in deliberation and immediately threw the case out, claiming that the accused were "denied the right of discovery"; prosecutors did not ask for an appeal.[45]

By 1966, the spirited movement to make the biological sciences more relevant and humane had grown to such an extent on all three Bay Area campuses that the idea was becoming not just acceptable, but accepted and natural. But at the same time, and despite at least modest signs of support, momentum actually showed signs of receding. Little coherent pattern could be detected in the unlikely mixture of policies, platforms, and ideas. The energy to make the biological sciences more responsive to human needs hid what had manifested on local campuses a few years earlier but was not so obvious at the time—that protesters who cared about bioscience research did not walk in lock-step.

Late in 1966, just around the time that the UCSF abortion trial was dismissed, Berkeley students staged a two-day teach-in to protest the inhumanity of the Vietnam War. Thousands of people paid rapt attention to the charismatic performance of such radical luminaries as Jerry Rubin, the pediatrician and author Benjamin Spock, the socialist leader Norman Thomas, SNCC leader Bob Moses, the novelist Norman Mailer, and the immensely popular local editorialist I. F. Stone. As the formal event came to a conclusion, leaders in the Berkeley chapter of Science Students for Social Responsibility distributed a pamphlet describing an academic program at Berkeley that "secretly conducted research on biological and chemical weapons." To publicize the "co-optation of Berkeley scientists" by the Defense Department, the activists called on their fellow students to join in a march to the nearby Oakland Army Base to protest Berkeley's activities as the primary accomplice of the military's massive chemical and biological weapons research program.[46]

Called the International Days of Protest, newspapers reported that thousands of students marched from Berkeley to nearby Oakland. The spontaneous demonstration generated enormous publicity; they picketed and sang songs of freedom, barricaded the entrance to the base, and sat on military base railroad tracks only to escape at the last second from oncoming trains. Just as protesters settled in, the Oakland police arrived with billy clubs, dogs, and tear gas. March leaders, unwilling to challenge authority, immediately called on demonstrators to return to Berkeley. Virtually all of the protesters complied. Jerry Rubin, frustrated by the timidity of those leading the Responsible Science movement, declared, "a movement that isn't willing to risk injuries, even deaths, isn't worth shit" and pulled out. Many of his notorious comrades followed.[47]

Sensing there might be something new about the biological sciences, Stanford students made a serious effort to root the radicalism of lifesaving research in its own grievances. For this task they took ideas from wherever they could find them, but most of all from the antiwar movement. Stanford student and de facto spokesman Brian Pugh anticipated that "if there is going to be real change, to make sense, protest must take the form of destructive criticism of a destructive system." Like-minded students began organizing in May 1966 with enough meetings to compress a lifetime of politics into a single year. Episodic protests took place throughout 1967, culminating in an arm of SDS staging sit-down strikes that blocked the doorways of the university's Applied Electronics Laboratory and the off-campus Stanford Research Institute. Railing against what they saw as the deathly pallor of academic science, Stanford's activists truly thought they were exorcising demons from science. Then President Nixon selected as the U.S. Assistant Secretary of Defense Stanford Trustee David Packard, who also served as president of Hewlett-Packard and director of FMC and Chrysler, among others. To Stanford's active student body, Nixon's appointment screamed blatant conflict of interest—together, the federal government, corporate America, the military, higher education, science, and technology.[48]

On 3 April 1968, 1,500 students from fourteen protest organizations gathered in "Dink" Auditorium. Called A3M, or the date that the movement was supposedly born, Stanford activists demanded that the university immediately cease all "chemical and biological weapon, classified Vietnam related, and counter-insurgency research." They had settled in for what they hoped would be a grand spectacle of public protest. It soon became apparent, however, that A3M was going to be, at best, a modest disruption, one where the distribution of Vietnam newsletters, the chanting of antiwar slogans, and petitions for "the principles of openness in research" would constitute the radical fringe.[49]

The activists should hardly have been surprised when Stanford trustees, who had always responded to student protest with rational objection, rose to the bait again. On 8 April embattled and outgoing University President Wallace Sterling announced that he saw "no reason to speak to the demands or issues of the disruptors." Demonstrators who remembered the benighted "Dink event" were relieved that Sterling had gone public with his policy to ignore them. Sterling gave the students another opportunity.[50]

On May first, A3M staged a sit-in that took over the administrative and business offices in Encina Hall, but neither university administration nor the police responded. As Brian Pugh recalls, the recent ineffectiveness of A3M galvanized about 900 determined protestors to remain inside Encina Hall, "for as long as it would take to breath meaning into our ideals." So students settled in for the evening, watched the SDS-Resistance film, "Battle of Algiers," and then passed a series of guidelines for all future protests: "there will be no violence against people, no destruction of property, no breaking into classified files." The students' high-minded ideals of nonviolence, just as those of university administration to ignore the protesters, would not last through the next morning. When the sun rose on May 2, university police had already surrounded Encina Hall; by 9:00 A.M. there were broken noses, broken bones, and many broken windows and walls. Throughout the police took blotter-photos to identify the individuals who had destroyed property, but the cameras were broken and the film, already developed at the Palo Alto City Police Station, somehow disappeared.

In the face of everyday terror that seemed to have taken over the campus, Stanford trustees nudged President Sterling into a more aggressive posture by giving him "emergency powers." Then, instead of confronting the students, the trustees outflanked them by putting the Stanford Research Institute (SRI) up for sale. Stanford students were angry and confused: they wanted to redirect war-related research into life-related research, but they could not influence the right kind of research if SRI became private and independent. So protesters shifted toward a new strategy: protect SRI as a legitimate university research facility.

In the early morning of 15 May, thousands of students boycotted classes and descended on Page Mill Road and Hanover Street surrounding SRI. In a series of disappointing protests, this one started out just as dull as the others—most of the attention was directed at an ad hoc committee collecting money to support the Jeff Browning, Vic Lovell, and Jay Saunders' bail fund. But then, in the midst of the morning commute and the arrival of television camera crews, frustrated drivers began ramming blockades and driving through chains of protesters linked arm-in-arm. Many students retreated by climbing to the rooftop of SRI, while others found safety by breaking into the SRI buildings. Police fired tear gas, but the students threw some of the canisters back onto the streets and into the morning traffic. One unidentified individual jumped over a barricade, raced to the entrance of the main SRI building, and spray-painted in large letters on the facade of the front wall: "Research Life—Not Death." For a moment, it seemed to many of the students at SRI that A3M had found its stride, but in reality the future held perils beyond even their idealized reckoning.[51]

By the end of the 1960s, all of the factions strutted about as self-appointed vanguards in search of followers: the SHO, MRU, A3M, the Diggers, Scientific Workers for Social Action, Environmental Action for Survival, Science for the People, Science for Public Death, and so on. But few could see common purpose among the many activists calling for practical and humane bioscience research. On the surface, the schisms split those who opposed science from those who believed in its uplifting potential. Indeed, along the antiscience path walked groups like the Diggers and Science for Public Death who appreciated the teachings of Henry David Thoreau, admired the photography of Ansel Adams and John Muir, and rejected the peaceable writings of Rachel Carson. They preached the interconnectedness of life, including humans in nature, and called for the coexistence of humans with nature. On the pro-science side were the activists in such organizations as Science for People or Scientists for Social Responsibility, who embraced the ideal that science—and scientists—had a obligation to serve society. But even these two worlds were not so neatly divided. Indeed, within the antiscience activists was a coalition of social-gospelers who wanted to leave science to God, while on the pro-science side were scientific ecologists who advocated organic science—natural, rather than invasive, experimental techniques.[52]

Despite all of the competing factions, however, the crucible of the movement was the pro-science activists, simply because they were more likely to end up in the laboratories while the antiscience groups were not. They recognized the necessity of fundamental research, but with varying degrees of restriction and regulation, from orthodox budget-cutting to expanded sponsorship of applied bioscience, from voluntary controls on individual experiments to government-supervised research regulations, from protection of patients to universal medical care. They did not want to change the entire system; more often they meant to provide palliative change obliquely related to some of the more revolutionary ideas promoted by their more radicalized brethren. Bioscience humanists opposed Berkeley's CBW research program, wanted to redirect the work conducted at Stanford's SRI, and supported UCSF's unofficial abortion clinic, because these kinds of efforts symbolized their conception of what was right about the biological sciences. They did not want the public directing the shape of bioscience research, the patient at the leading edge of a revolutionary social movement, or the creation of an entirely new biomedical establishment; rather, they preferred that research, health care, and medicine remain a personal matter. In general, they understood the biosciences in rigidly circumscribed terms, and preferred to rely on experts—not coincidentally, whom most of the protesters themselves hoped to become—to reshape bioscience research so that it had more purpose, more utility.

Amid the chaos and divisions within the ranks, the whole of the movement sparked an impulse to change. Ultimately, the desire for a conscience in science, rather than a new consciousness, drew into the movement the general public as well as bioscientists in academia, such as biochemist Leonard Herzenberg or biologist Donald Kennedy, because everyone could support the effort to make science more responsive. Causes, such as the effort to halt chemical and biological weapons research, could be joined without anyone risking professional careers. In voicing a desire for more practical application of bioscience research, protesters advocated an ideal that promoted the interests of all humanity, not one that tried to aid one group over another. Moreover, this message had popular appeal because many activists had carefully distanced themselves from the radical social protests around them. The audience may not have embraced the whole movement, or rejoiced in its ascendance, but they listened to the message and in some cases tolerated the messenger. Ironically, it was these bioscience humanists, although occasionally unpopular for their imperialist tendencies in the 1960s, who would eventually become the handmaiden for a commercial bioscience industry in the coming decade.

Chapter 6

A Season of Policy Reform

How can one ask the public to provide support, much less facilities, for the intellectual gratification of one select group? . . . The answer, of course, is simply one cannot. As long as a group is dependent upon public support it must seek some means of contact with the values of the enveloping society, and the moment it does this it departs in some measure from the ideal purity.

—Charles Sanders, 1967

The making of a biotechnology industry involved both the loosening of the scientific research agenda through conflict between pure and applied researchers as well as a counter culture that raised challenging questions about the value of pure research. Indeed, the precondition for the coming revolution was this rare combination in which a splintered field could not contain popular concerns that had crystallized around a desire to make the biological sciences more responsive to human needs.

However, behind the discipline's unstable structure and the protesters' novel vision sits a rare event in the history of science: a radical shift in state research policy that was equally significant in pushing the biological sciences in new directions. Critiques of pure research may have taken hold and spread at the grassroots in the Bay Area, but the policy changes that pushed the biological sciences in new directions took place in Washington, D.C. This chapter is, therefore, a study of policymakers, their interactions with protesters and leading investigators, their motives, and their ultimate decisions. In no way does the shift in science policy suggest that all biologists underwent some profound ideological conversion. Indeed, the broader impact of new science policy cannot be readily discerned without a detailed examination of how investigators reflected upon and responded to new opportunities, which is the focus of the next chapter. But to understand the rise of a biotechnology industry requires an understanding of how, beginning in the late 1960s, policymakers upset the old model of patronage and set in place a motive to pursue new experimental directions.

War on Poverty, War in Vietnam, and War on Basic Bioscience Research

From 1946 until the mid-1960s, Congress never seriously questioned the importance of science, nor was it ever a matter of politics. Obdurate though bipartisan support remained, it weakened for a brief moment in the early 1960s when a series of scientific mishaps—thalidomide, fluoridation, radiation, the Cutter Labs polio outbreak, ecological damage described in *Silent Spring*, and so on—sensitized the public to a dark side of science. In response to these catastrophes and others like them, the populist Senator Estes Kefauver of Tennessee submitted a bill that would amend the Food, Drug, and Cosmetic Act of 1938, proposing greater protection for consumers of pharmaceutical drugs.[1]

Kefauver's bill was, from the outset, a magnet for controversy. It was a single, integrated federal program that ignored the time-worn principle of scientific autonomy. Simply put, Kefauver envisioned greater regulatory powers for the FDA to inspect the manufacture of drugs. Before Kefauver submitted his bill, the FDA merely required proof of a pharmaceutical's "relevant toxicity"—that the drug in question was safe and would not harm the consumer—a compulsory requirement that could be met with a few relatively straightforward experiments. Kefauver's bill added an additional requirement—"proof of efficacy"—that the drug was both nontoxic *and* effective in the treatment of the targeted ailment. As the public grew more concerned about science, amendments to Senator Kefauver's original bill grew more purposeful.[2]

Kefauver's bills provoked criticism among scientists who hated the idea of government intervention. When *Science* asked biologists from eighty-one colleges and universities what they thought about stronger FDA regulation, all eighty-one voiced serious reservations. One worried that new federal regulations might cause an "unconscionable growth of paper work" and turn biological research into "a bureaucratic monstrosity." Another feared that biological research "would simply cost twice as much to carry out." No less disheartening, *Science* provided testimonial evidence that Kefauver's bill would fall with especially sharp brutality on drug developers. An FDA preliminary report confirmed the wild anxieties: drug developers would have to provide evidence of chemical purity, qualitative evidence that all experiments adhered to "good laboratory practices," and that drug compounds had no toxic side-effects on laboratory animals. Investigators would have to design elaborate tests for human subjects; collaborate with physicians willing to submit their patients to testing; identify a large, random, and anonymous experimental pool of patients; and provide detailed documentation of "drug balance of risk to benefits" ratios for human use. The entire sequence would have to be repeated to identify "appropriate dosage." Finally, investigators would have to prepare double-blind tests carried out by physicians at multiple sites, none of whom would be aware that they were testing the clinical effectiveness of the drug as it compared with the effectiveness of a placebo. In all phases, FDA officials would need time to review documents, could request a retest for any particular phase, or ask applicants to redo the entire process again.[3]

Under such constraints, most of the bioscience community looked upon Kefauver's bill as constraining drug development. Doubly ironic, in order to prove the purity and efficacy of a new drug, investigators would need a deeper understanding of the molecular basis of the disease in question; to know more about the mechanism of a drug required greater fundamental knowledge about the molecular biology of the disease and the chemistry of the product. From a scientific perspective, Kefauver's idealistic attempt to protect consumer health would, in practice, reduce medical research, raise medical costs, and emphasize pure research rather than medical care.

Despite the inherent problems with Kefauver's bill, the public had become so worried about scientific catastrophe that the main political question at the time was not whether the bill would pass, but the speed with which it would move through Congress. No formidable political opponent spoke against the bill; Kefauver and his fellow sponsors avoided the delay of Senate-first strategy and pushed the bill simultaneously through both chambers. The Senate voted unanimously in its favor. As Kefauver's bill entered the House for a floor vote, President Kennedy, an unwavering defender of the scientific establishment, caught wind of shifting popular opinion and recommended that the House approve the measure to "protect our consumers from the careless and unscrupulous." On 10 October 1962, President Kennedy signed the Kefauver-Harris Amendment, calling it a "noble" bipartisan effort, and awarded Frances Kelsey, a drug-review official at the FDA, the Distinguished Federal Civilian Award for single-handedly halting further distribution of thalidomide.[4]

Kefauver's amendment to the FDA Act strengthened a modest and circumscribed agency to police a growing problem in pharmaceutical research, production, and distribution, yet it scarcely satisfied the public. The FDA's new and stronger regulatory powers still relied on scientists to police themselves and to protect consumers, and focused on the prevention of scientific mishaps through regulation—not redirection—of runaway research. Furthermore, the same Congress that voted overwhelmingly in favor of stiffer FDA regulations also voted to increase the NIH budget from $430 million to almost $1 billion by the time the new regulations went into effect in 1964—and chose to ignore the fact that years of unrelenting 15 to 25 percent growth had already raised funding for the biological sciences to untenable levels.[5]

Beyond the passionate pleas for stronger FDA authority, Congressional emphasis on regulation rather than redirection located both Democrats and Republicans toward the political center, while voters on the political left and right continued to worry that federal support of basic bioscience research had ignored more pressing health or fiscal concerns. And yet, even though the new FDA regulations may not have satisfied voters, it was nevertheless a logical legislative strategy at the time. The public may have been frustrated with fiscal abuse, or misuse of federally funded scientific research, but in the early 1960s, their frustrations lacked focus, and the nascent core of the emerging protest culture lacked a coherent vision or direction. No one at the time knew how long protesters would care about the issues, or the ferocity with which activists would attempt to affect change. As scientific mishaps seemed to die down by late 1964, so too did legislators' desire to redirect bioscience research, becoming less pronounced than it had been just one year earlier, or would later become. While bioscientists looked upon the new FDA regulations as scientific obstruction, and while Congress considered it scientific reform, a mighty host of critics in the public realm continued to throatily assail the scientific status quo.

The cautious, compromising approach of bioscientists and Congress—and their willingness to ignore public concern—allowed plenty of room for less-than-conventional politicians to seize the issue, and few during the mid-1960s could claim to be more responsive—or more unconventional—than Louisiana Senator Russell B. Long, son of an original populist, Huey Pierce Long. Much like his father, Russell Long was a shrewd operator in Congress who spoke zealously for the interests of the powerless by waging verbal war on elites. Long-style politics may have attracted a core constituency in Louisiana, but it generally offended most everyone else; for most of his political career, America simply ignored Russell Long. But by the mid-1960s, much had changed, and Long's stubborn contempt for academic "high hats" and incessant attacks on the "ivory tower" mirrored the nation's growing distrust of academia and the public's nascent desire to make science more responsive and responsible.

In spring 1965, Russell Long seized the Senate floor and with reckless abandon launched into a detailed discussion of a simple blood test given to newborns to detect the presence of phenylketonuria (PKU)—a metabolic disorder known to cause some forms of mental retardation. Long's unusual grasp of complex scientific principles shocked his colleagues, as did his sympathetic discussion of how doctors could use the test to prevent fetal brain damage. According to Long, the NIH–Public Health Service paid $251,700 to three researchers to develop a standard PKU test, only to watch helplessly as the three filed for a patent for their invention and then entered into an agreement with a private company to manufacture and sell the test kit for a profit. Long boldly pressed the attack: before the patent, the test cost $6; with privatization, the test cost $262. Long used the case of the PKU test kit to illustrate several of his concerns: academic researchers are self-serving and driven by profit, bioscientists withhold critical information vital for improved medical care, and, most importantly to Long, "publicly financed university research is contrary to the public interest." Though many people sympathized with Long and his charges, his political extremism, much like the activism of radical student protesters, would probably have had little lasting effect had it not been for another powerful figure who also questioned federal support of pure bioscience research: the inveterate President Lyndon B. Johnson.[6]

A master of consensus, contemporaries once admired Johnson's instinct for compromise. But it is a standard rule in American politics that while the quest for consensus may minimize risk, it also brings about centrist accommodation, and by 1966, America felt it needed something more. Indeed, Vietnam, the War on Poverty, student protest on university campuses, inner-city violence, and racial integration all pushed the country in opposite directions, stretching thin Johnson's precarious political coalition. Moreover, the steady growth of the nation's economy—which had been going on since Kennedy's term in office—had taken an abrupt turn: inflation and unemployment levels were rising, real wages and consumer spending were falling, and the nation's GNP was stagnant. Johnson worried that political whiplash would destroy his reelection campaign. The president's economic advisers recommended that he slash spending to choke off inflation and regain control of the $7-billion deficit, the largest in years. Many wondered where spending cuts would come from—Republicans wanted to cut social welfare programs while Democrats wanted to raise taxes—but budget cuts and taxes had already reached what Johnson's advisors told him were dangerous levels. Thus, the political circumstances in 1966 were very different than just a few years earlier.[7]

Against this uneasy backdrop, comments by two unassuming White House aides—science adviser Donald F. Hornig and his deputy, Ivan Bennett, Jr., former chief of pathology at Johns Hopkins—became key political strategy for an increasingly desperate Johnson administration. Their contributions came about somewhat unexpectedly, as an informal presentation to a group of biologists, as a plea to recognize political reality and to scale back their request for federal patronage. Subtle and reasonable they may have been, Hornig and Bennett asked bioscientists to engage the profound ideological shift taking place across America, a considerable somersault from a notably conservative scientific community that had come to expect complete deference from policymakers. For instance, just six years earlier President Eisenhower's science adviser ambitiously proclaimed: "it is not possible to assign relative priorities to various fields of science," which apparently justified unlimited spending in all scientific fields. Or, the bold statement by Kennedy's science adviser four years later: "What is it that should determine our national budget for basic research in universities . . . ? Just one thing, I submit. It is . . . when, and only when, every competent research scholar in our universities receives adequate support." This awesome sentiment had been the crux of an elaborate argument for almost two decades, and an ingenious one, for it worked far better than anyone could have imagined. But Hornig and Bennett had noticed that the apparently well-intentioned proposition that fundamental research deserved full support concealed some explosive political dynamite for their own boss, and they believed it required prompt attention if they were to protect their administration.[8]

Amid Johnson's deepening political crisis, Bennett carefully selected the annual meeting of the Federation of Americans Societies for Experimental Biology to challenge bioscientists' naive perception of political reality, and suggested that they begin rethinking the value of their work to society: "The fundamental premise that we need more money for research grants, for training grants, and for the physical facilities . . . so permeates our thinking and our way of life that there seems to be something contrived and artificial about any situation that calls for justifying the view that basic research is in the national interest." Bennett challenged the experimental biologists gathered at the conference with difficult questions about the time-honored, academic viewpoint that, as a matter of foreordained right, all scientists deserve federal money to support fundamental research:

> As impressive as such ringing statements may have been to legislators and appropriators in the halcyon days of yesteryear, they are now regarded not as expert testimony, but as special pleading, which—and it is time we admitted it to ourselves—is exactly what they are. . . . While all of us believe that planning of science should be our responsibility and ours alone, most of our justifications for support . . . have taken the form of saying that we should continue to do exactly what we have been doing—only more of it.

Without committing the Johnson administration too much, Bennett concluded his presentation with a forewarning: "It is abundantly clear that if we don't take the lead in jettisoning some of the excess baggage, others will."[9]

At about the same time, Donald Hornig delivered a similar message to another group of scientists, but with a more direct warning: "The scientific community is going to have to learn to articulate its hopes, to describe the opportunities which are before us for practical advance, to express the excitement of the new intellectual thrusts—but to do these in terms which the American people, who are expected to pay the bill, will gradually understand and have faith in. There is no other alternative." Hornig and Bennett's critique of fundamental research in the biological sciences echoed public and legislative concerns, and more important, signaled a deeper awareness within the White House that the relationship between fiscal responsibility, the isolation of the biological sciences from the public, and popular opposition to scientific elitism was a potent combination that held tremendous moral authority among voters.[10]

Johnson's policy strategists—including Bill Moyers, a fellow Texan who had joined the Johnson staff when he was vice president, and George Reedy, Johnson's general-purpose press aide who spent most of his days checking shifting political interests—instantly recognized that the combination of ideas expressed by Hornig and Bennett could serve as a powerful counterforce to the administration's weakening political stature. In the president's Special Message to Congress in January 1966, Moyers tucked away a subtle jab at fundamental bioscience research and a more pointed call for change. The speech can hardly be described as a great state paper; it suffers from a long list of unrelated issues and concerns that together show little coherent pattern. Still, Johnson's optimistic appraisal of bioscience's potential reflected the surging mood of the time. "We must make sure that no lifesaving discovery is locked up in the scientific laboratory," intoned Johnson. "The day of the great discoverer is over. Our task now is not discovery, or exploitation of scientific laws, or necessarily producing more knowledge. It is the soberer, less dramatic business of distributing money to more humane scientific research projects, of seeking to re-establish public health, and of meeting the problem of disease in urban centers."[11]

Johnson continued to test these themes at other speaking engagements. For instance, two months later, at the American Association of Medical Colleges' annual meeting, Johnson endorsed a more rigorous commitment to improving medical care: "Presidents . . . need to show more interest in what the specific results of research are—a great deal of basic research has been done . . . but I think the time has come to zero in on the targets—by trying to get our knowledge fully applied." Douglass Cater, special assistant to Johnson and the point of access to the White House for those seeking federal support for bioscience research, recalled the "interesting playback" on Johnson's call for biomedical application, where almost "all legislators—Democrats as well as Republicans" gave it "practically unanimous recommendation." Though applied bioscience research would never serve as the centerpiece for Johnson's new political agenda, it drew overwhelming popular support for his new moral agenda. Not surprisingly, the scientific community arrived at an altogether different conclusion: "The President's initiative . . . caused an explosion among the scientists and in the universities," commented one concerned observer. "Support for applied research and development was to be substituted for support for basic research by an anti-intellectual, unsophisticated President who could never understand such things."[12]

The biological sciences, still stinging from Johnson's rebuke, received another devastating blow a few months later when the Department of Defense (DOD) distributed an internal study, "Project Hindsight." Analyzing retrospectively the development of twenty important military weapons, the authors of the report observed that the contributions of university research was minimal, the scientists who contributed the most were "mission-oriented," and the lag between discovery and final application was shortest when scientist bypassed fundamental questions and worked directly on subjects targeted by the sponsor. The study concluded that the DOD's support of basic research—which had grown 30 percent annually from 1946 to 1956, and then continued at a rate of 15 percent between 1956 and 1966—resulted in few, if any, useful developments.[13]

The president's actions and the DOD's report "fell like a bombshell on NIH officials," who then immediately asked RAND Corporation to conduct an internal assessment of their own investment in fundamental research. After three months of rigorous investigations, including numerous on-site visits at UCSF and Stanford, RAND officials concluded that "clinics are not related to the [problems faced by] research programs." RAND determined, in short, three problems endemic to the biological sciences: first, there was a profound "lack of proper incentives in the management" of grant money and that federal money had been distributed carelessly by agency representatives; two, university administrators frequently "used surpluses to cover the deficits in other [departments]," which in practice helped sustain underperforming, unwanted, or irrelevant experimental programs; and three, there was no evidence of "accountability of department expenses." Deeply concerned about these initial findings, NIH officials asked RAND to conduct another follow-up study to confirm the opinion of the president and Congress, the DOD report, and the first RAND investigation, and again, RAND determined that while "the involvement of medical schools in research has expanded rapidly since World War II . . . education and patient care outputs of the medical schools" suffered because the NIH had inadequate "management of research" and ignored the inefficient "allocation of resources" by grant recipients. At this moment, the biological sciences confronted something far more serious than a passing crisis, but an episode that revealed deeply rooted disciplinary and administrative problems.[14]

For Johnson, "Project Hindsight" and RAND's conclusions confirmed what he long suspected, and he spent the next six months hammering away at what one writer called "the great research boondoggle." With words freighted with importance for his War on Poverty, Johnson began articulating in summer 1966 a new conception for biology, directed again by an activist state, but this time one that would restrict what investigators had traditionally enjoyed: experimental autonomy, federal patronage, and isolation from public concerns. For instance, on 27 June 1966, Johnson told an assembly of NIH directors that "too much energy was being spent on basic research and not enough on translating laboratory findings into tangible benefits for the American people." Soon after he asked the Surgeon General and other NIH directorates to review experimental objectives and, if necessary, reshape them to get maximum results from these projects. He also suggested to scientific audiences that he might ask Congress to look into cutting basic research funds. In his desperate effort to rebuild political consensus, Johnson's messages to the biological sciences had triumphant political consequences: liberals were convinced that a dramatic shift in state science policy toward practical application would have a positive effect on medical care; and anti–New Deal political conservatives believed that the federal government played too great a role in university research and that continued expansion of federal research patronage was fiscally irresponsible.[15]

Johnson's political logic seemed unassailable, and for a brief time, it seemed as this single issue—the application of research in the biological sciences toward human needs—might become a defining feature for his legislative agenda, just as it could have also become a critical theme for dissatisfied voters. The president, populist legislators such as Estes Kefauver and Russell Long, and student activists may not have spoken in unison on this issue, but when they did speak, they did so powerfully, and their efforts to reshape the biological sciences did not go unnoticed by other politicians. California Governor Edmund G. Brown reported that on the basis of casual observation, research conducted on space and defense should be redirected toward solving much more pressing problems of smog, traffic, water shortage, or sewage and waste disposal. At

Stanford, Senator Al Gore told students that he supported their efforts to end the military use of napalm, while Robert Kennedy supported MACV's attempt to "provide blood to those who needed it," even if for the North Vietnamese. Even the conservative Republican Barry Goldwater admitted to a stunned Stanford audience that he appreciated any attempt to rein in swollen federal budgets, even if it meant cutting fundamental research sponsored by military agencies. Corporate executives also joined in, such as Motorola Chairman Robert Galvin, who took out a series of full-page ads in Stanford and Berkeley student newspapers to discuss how his company did much more than contribute to the development of napalm, but that it also conducted research that made positive contributions to humanity. Carl Djerassi, cofounder of Syntex, one of the nation's largest manufacturers of pharmaceuticals, vigorously campaigned to have company headquarters moved from Mexico City to Palo Alto to take advantage of, among other things, the Bay Area's new appreciation for applied bioscience research. Even Pope Paul chastised scientists that basic research "divorced from the higher interests of man" had become "sterile, useless and, let me say it, harmful": one observer at the UCSF Medical Center commented that he and his colleagues "breathed a heavy sigh of relief that the Pope did not sit on grant review boards."[16]

Compounding the biosciences' general unease and precipitous decline still further was the loss of their most ardent defenders, also in 1967: the retirement of NIH director James Shannon, who had capably led his agency through its greatest periods of growth, and the NIH angel investor on Capitol Hill, Congressman John Fogarty of Rhode Island, to a fatal heart attack. As chairman of the subcommittee handling NIH appropriations, Fogarty managed to protect the NIH from criticism and secure generous increases in the NIH budget, consistently well above the administration's requests. Moreover, the NIH's Congressional subcommittee found itself depleted of several key supporters due to electoral defeats in the 1966 elections, most of whom were replaced in 1967 by newcomers who saw the biological sciences as prime for remaking.[17]

It was still possible to argue in 1967 that basic research was as productive as it had ever been, and that the scientific community remained essentially intact. To illustrate the paradox of incredible fundamental research achievement and declining public support, consider the circumstances that Stanford biochemist and Nobel laureate Arthur Kornberg encountered in 1967. Much like the previous twelve years, Kornberg had an enormous budget, the best equipment in the world, and a world-class staff, all dedicated to accomplishing perhaps the single most ambitious biological question of the age: could scientists artificially create a biologically active piece of DNA in vitro?

In the race to duplicate nature's processes, Kornberg and his colleagues had the inside track: his own lab had experimental evidence that suggested that it was possible to do so, and his staff had recently developed a remarkable ability to manipulate polymerase—the enzyme that DNA uses to duplicate itself—to manually and artificially assemble in proper order DNA bases to make an exact replica of a DNA imprint. For years, Kornberg and his laboratory assistants had used DNA polymerase to assemble a variety of DNA chains of simple bacteria such as pheumococcus, hemophilus, and bacillus subtilis, and each time they had successfully produced an exact copy of DNA that was identical in both form and composition to the original, but never biologically active. Then, in early 1967, several participants in Arthur Kornberg's laboratory—notably, the postdoc Mehran Goulian and visiting professor Robert Sinsheimer—had some experimental success with a tiny bacterial virus found in the Parisian sewage system called Phi-X-174 and a newly discovered enzyme called, appropriately enough, DNA-ligase that they found could "artificially resuscitate" or heal broken DNA chains. However, other laboratories were having similar successes with the Phi-X-174 bacteria and DNA-ligase. But the outcome of the race was largely determined by Kornberg's previous experience and expertise with polymerase, as his team of researchers quickly used the enzyme to assemble a single strand of Phi-X-174's 5,000-nucleotide chain that was identical in form and composition to the natural virus, and then sealed the new strand with the DNA-ligase enzyme. Immediately, the synthetic piece of Phi-X-174 began reproducing its identical progeny, a fundamental action typically associated with life's most basic processes.[18]

It was nothing short of a miraculous scientific achievement, but Kornberg recognized that the public no longer cared as much for this sort of work. Although he had always been much more comfortable in his laboratory than in the public eye, Kornberg reluctantly tried to promote his discovery and use it imaginatively to regain popular support for fundamental research. Breaking dramatically from tradition and from his own personal comfort zone, Kornberg announced the discovery to the press rather than publishing his results as an essay in a peer-reviewed scientific journal.[19]

The public approach paid immediate dividends. Kornberg's announcement made headlines across the nation and generated a buzz among people normally not interested in science. According to reporters, the replication of DNA stood as one of the greatest bioscience achievements ever; one newspaper went so far as to describe it as the single greatest bioscience news story of the twentieth century, ahead of DNA, the eradication of polio, and the first successful artificial heart transplant. Hyperbole aside, almost everyone agreed that Kornberg deserved an unprecedented second Nobel Prize. Kornberg's scientific achievement was pivotal in another respect. This "victory for pure research," as he called it, had no obvious practical application; it was, in his mind, another fundamental breakthrough. Kornberg and the Stanford Medical Center's news bureau went to great lengths to ensure that newspapers described the experiment as "pure research," and they deleted any popular description that implied practical use, such as "the creation of life in a test tube." Reporters must have been sorely disappointed when told that a more accurate account of the experiment would be "synthesis of the inner core of a virus."[20]

But Kornberg's moment of persuasion was pitifully brief, and not just because he had made the fateful decision to reach out to the public. Days after Kornberg announced his discovery, President Lyndon Johnson broke from his carefully prepared remarks at a Smithsonian Institution bicentennial convocation in Washington, D.C., to offer an altogether different interpretation. Whether an attempt to dismantle biology's semifeudal caste system, upstage basic research, or divert attention away from his failures in other political arenas, Johnson told the press about an amazing discovery by a biologist at Stanford. "What are you going to read about tomorrow morning?" asked Johnson. "It is going to be one of the most important stories that you ever read, your Daddy ever read, or your Grandpappy ever read." Then, defying Kornberg's plea for scientific accuracy, Johnson intoned: "Some geniuses at Stanford University have created life in a test tube!"[21]

President Johnson's declaration seemed at the time a master political stroke. In a sense, Johnson's redefinition of Kornberg's fundamental research neatly reflected his own conception of problems with the biological sciences and perfectly suited the nation's desire for greater attention toward medical research. The press, appreciating President Johnson's more sensational rendition of the experiment than Kornberg's, provided exactly the kind of story that its audience wanted to read: the *Los Angeles Times* declared that the discovery "may open up new avenues of research in finding out what takes place when normal cells are changed into malignant, cancerous cells"; the usually sedate *New York Times* described the work "as breathtaking as those exposed several decades ago." Local papers were even more effervescent: the *San Jose Mercury News* declared that this discovery "could be the first step toward the future control of certain types of cancer"; not to be outdone, the *San Francisco Examiner* reported that "one of mankind's most impossible dreams has been the creation of life in a test tube. Last week, scientists moved a step closer to making the dream possible."[22]

Kornberg's inability to define the meaning of his truly incredible discovery is a singular example of how far the shadow of public frustration had fallen on the biological sciences in 1967, especially in the Bay Area. Local bioscience laboratories continued to churn out incredible discoveries, just as they had in years past. But the public saw scientific autonomy as merely an attempt to shroud scientific intentions behind thinly veiled promises of everlasting life. In a sense, the habit of isolation had robbed them of indispensable public trust that might have come had investigators engaged in or contributed to a sense of democratic participation before 1967. Kornberg's decision to use the press to announce a major bioscience breakthrough served as a historical pivot in another respect: future generations of bioscience investigators would happily follow this same approach, but with far greater professional and financial returns. But in 1967, Kornberg and his colleagues discerned this fateful outcome as merely another instance in which high scientific principle had been sacrificed to accommodating popular concerns.

Out of the fevered, chaotic, initially defensive and ultimately compromised setting came a wave of verbal attacks that rained down from leaders in both houses of Congress and from both parties. They aimed their most lethal condemnations at scientists who continued to submit requests for increased funding. On the right, Republicans such as Representative Gerald Ford, House minority leader from Michigan, preached fiscal responsibility to sponsoring agencies such as the NIH. Republican representative from Ohio Frank Bow, once a proponent of funding increases for basic bioscience research sponsored by the NIH, reversed course and informed his colleagues that "research spending is a prime area for economy." Other economizers, such as Representative James Fulton, ranking Republican on the Science and Astronautics Committee, and Leslie Arenda, the minority whip from Illinois, wanted to cut nonmilitary and nonspace-oriented basic research budgets—in particular, any public money allocated to the biological sciences through the NIH.[23]

Despite the sincere opposition that fiscal conservatives felt toward inflated bioscience research budgets, disillusionment with the field ran deepest on the left. Democrat Senator Proxmire—later author of the infamous "Golden Fleece" awards—pointed to the NIH in particular as one of the "worst offenders" of publicly supported research, dragging out carefully chosen examples of misused funds such as "A Social History of French Medicine, 1789–1815" and the enormous amount of time and money dedicated to "revising pickle standards." Democratic Senator Fred Harris of Oklahoma also saw basic research as a "political boondoggle" that directed taxpayer dollars out of poorer states like his own and toward a few select regions, especially the three research universities concentrated in California's Bay Area. Liberal Democrats such as Senator Ted Kennedy of Massachusetts and classic New Dealers such as Vice President Hubert Humphrey of Minnesota also chafed at the thought that fundamental bioscience research might get funding they wanted for public health, antipoverty programs, and education. Congressman L. H. Fountain of North Carolina, whose sharp criticisms of the biosciences had been ignored only five years earlier, found himself in the national spotlight.[24]

Johnson's search for consensus made for strange political bedfellows, as Republican and Democratic representatives, who collectively agreed that a shift from basic to applied bioscience research was both morally and fiscally responsible, prepared to abandon the scientific community en masse. Science writer Elizabeth Brenner Drew described this powerful political arrangement as a group alliance rather than the effort of a few key individuals. Indeed, it was an unusual combination of political forces that had rallied around this single theme: Republican fiscal conservatives and moderate Republicans such as Thomas Kuchel of California had joined powerful Senate liberals such as Hubert Humphrey and a retreating Southern bloc that was desperately looking for an issue that might deflect public attention away from concerns about civil rights.[25]

The frenzied attack peaked in late 1967 when the bipartisan House Government Operations Committee issued what one scientific journal described as "one of the bitterest critiques a congressional group had ever directed at a federal research agency." The committee report charged the NIH with a thick catalog of failures ranging from "weak and ineffective central management" to administrative procedures that are "irresponsible, unscientific, and contrary to the best interests of the . . . community and the government." Moreover, this committee questioned the quality of research supported by the NIH, accusing the agency of favoritism in the distribution of money and "singlemindedly overfeeding basic bioscience research" to the detriment of teaching and medical service. Dredging up one polemical example after another, perhaps the most damaging—and most telling to the collective mindset of Congress—was the unwillingness of the NIH to implement the Health Science Advancement Award (HSAA), which was originally established in 1965 to provide relatively sparse $1-million grants to help develop "new and [stronger] health science activities." While NIH generously funded fundamental research programs throughout 1965 and 1966, just as they had in the past, the NIH apparently downplayed the HSAA in an effort to avoid being "flooded under with 15 applications or so." Officials at the NIH may have been pleased when only three applications trickled in during a two-year period, but Congressmen interested in health care and the cure and treatment of disease were certainly not.[26]

To address all of the perceived infamies committed by bioscientists pursuing pure research, Congress called for a strict subcommittee review of governmental policy by establishing the National Commission on Health Science and Society (NCHSS). All told, the NCHSS asked thirteen leading figures from various medical schools around the country to share their opinions about federal support of bioscience research; in a sure sign of Congressional opinion on these matters, the subcommittee only invited two investigators friendly to fundamental research. Not coincidentally, both were from Stanford: Arthur Kornberg in biochemistry and Joshua Lederberg in genetics.[27]

On the first day, presidents from four different medical schools described for their captive audience in painstaking detail the plight of clinical care and their inability to do anything about the current healthcare crisis; one perceptive administrator noted a class-like distinction between enormously wealthy bioscience research laboratories and understaffed, underfunded, and underappreciated hospitals. After hours of testimony, an obviously moved committee wrung their hands in anxious frustration, lying in wait to establish a measure of equilibrium between basic research and clinical care.[28]

It was then Arthur Kornberg's turn to testify. He opened with an innocent defense of basic research and a brief review of recent fundamental discoveries in the field, and then he sat back and waited to engage in an intelligent discussion of bioscience research in the modern age. Senator Ribicoff, on the other hand, assailed Kornberg much like a defense lawyer would crossexamine a hostile witness. After going through the formalities of general introductions, Ribicoff peppered Kornberg with a Socratic inquiry into his understanding of federal funding of research. "Is most of the research in the field being financed by the Federal Government at the present time?" asked Senator Ribicoff. Kornberg replied: "as a guess, I would say over 90 percent." Senator Ribicoff asked what were the main objectives of research, to which Kornberg replied that there was no ultimate goal because no scientist could objectively conduct research if he had a desired outcome in mind. Senator Ribicoff pressed him to predict some future practical health benefits that might result from his research, but Kornberg stubbornly refused: "I really do not have the capacity to answer some of the questions that are not well-defined and still so distant. I have learned from experience that it is more meaningful for me to focus on problems that confront me directly." The senator then delivered a devastating blow:

> Does a gentleman like you ever undertake . . . soul-searching as to the consequences that may come from breakthroughs and achievements in this field? You

talk about a democratic society . . . but the question that concerns [us] is science amoral? Does science concern itself with the ethical, social and human consequences of its acts and its achievements? How do we involve society and how do we involve the scientist in a humane objective in which people can live a decent, well-rounded life . . . ? At what stage does the scientist become concerned with the good as well as just the success of the work that he is doing?[29]

Kornberg immediately backtracked from his earlier statements by offering alternative interpretations and tried to divert attention to the ethical dimensions of the Senator's own work, such as his support of the Vietnam War. Senator Ribicoff would have none of it. Kornberg had given the senator his proverbial head. But the Senator chose to defer, graciously allowing his colleague, Senator Walter Mondale, the opportunity to deliver the fatal blow.

Mondale wasted no time. He told Kornberg that he found it "remarkable" that this was the first time Kornberg had ever testified before a congressional committee. "Is it any wonder," asked an incredulous Mondale, "that you are having financing problems? Is it any wonder that the public is not responding to its Congress to give you the funding that is reasonably needed to pursue your objectives of adequate research more fully?" Then Mondale delivered the mortal wound prepared by Ribicoff: "I wonder if part of the problem of the lack of public support for your kind of research . . . might stem from this reluctance to carry on a dialog about the real human implications of your research. . . . Isn't part of the problem of the inadequacy of public support for the very work you are involved in, traceable in part to the fact that you have avoided the public?"

Realizing that he had created an impossible position to defend, Kornberg desperately tried to reverse course:

You must know the kind of creature you are dealing with in the scientific community. . . . The biochemist who deals with molecules cannot afford any time away from them. Today I am not in the laboratory, I do not know what is going on at the bench. Tomorrow I will be less able to cope with the identity and behavior of molecules. The more I am estranged from the laboratory, the less competent I am. . . . It is, ultimately, more than a little forbidding for us to cope with news media and forums that are so unfamiliar.

The two senators could not have conveyed their point any more clearly.[30]

Later that same day Joshua Lederberg also testified before the same committee and tripped over the same moral redux. Lederberg did not want to talk about science policy with Congress—or the public's concerns about the application of his research—and he suggested that the entire problem could be solved quite easily if bioscientists had better "control of the dissemination of scientific information." He also doubted that the public could truly appreciate his profession's work or its significance. Much like Kornberg, Lederberg wanted the public's money and then wanted to be left alone while he used it to pay for his laboratory experiments. Emblematic of the degree to which bioscientists held a bunker-like mentality, Kornberg and Lederberg received hundreds of letters of support from bioscientists around the world, almost all of whom appreciated their efforts, and offered a sympathetic appraisal that neither Congress nor the public was truly capable of appreciating what they considered was the "undeniable" importance of fundamental research in the biological sciences.[31]

Immediately following the NCHSS subcommittee hearings, Congressional opposition to basic bioscience research sprang forth from a variety of corners. An especially caustic attack came from Representative Wayne Hays, a key Democrat from Ohio, who denounced a series of research programs that he once supported. In what must have been an embarrassing turn of political allegiance for the scientific community, Hays launched into a dramatic populist tale about a certain policy he had on his farm in Belmont, Ohio: "We only save about two of the best bull calves for breeding purposes," explained Hays, "and the rest of them are made steers and eventually wind up in the butcher shop. And while I was riding around thinking about this, it occurred to me that . . . if I were President of the United States I could not think of a better present that I would like the Congress to give me than a $5 billion gold-plated castration knife." And with words that must have given Bay Area bioscientists pause, he warned, "and do not think I would not know where to cut." On that very day the House, apparently visualizing the implications of Hays's painful metaphor, voted to equip President Johnson with a gold-plated scalpel in which he could cut to the bone anything that would allow him to sustain, simultaneously, Vietnam, the War on Poverty, and other social reforms and still appear budget-conscious.[32]

As the president and Congress prepared to plunge its knife deepest into the least-armored parts of the budget, federal science agencies warned bioscientists in the Bay Area and elsewhere that they should brace for dramatic cutbacks in funding, some of which could take effect retroactively and appear in the 1967 fiscal budget:

- The National Institutes of Health "will be able to fund, on the average, only about 50 percent of the new applications which have been judged scientifically worthy of support. . . . Our assessment of our funding capability for the immediate future indicates a further lowering of the percentage of approved applications."[33]
- The National Heart Institute "can, at this time, fund 33 new fellow-

ships nationwide (as opposed to the hundreds that they had funded every year in the past) and will cut the number of research grants from the present 2081 to 1820, and traineeships from 1415 to 1340."[34]

- The National Science Foundation "will offer 47 fewer research grants in fiscal 1967"—the first cut in the history of the foundation.[35]
- Health, Education and Welfare (HEW) will "defer $208 million in spending in 1968. Moreover, most HEW agencies do not have sufficient funds . . . to fully support all worthwhile research projects and will, therefore, postpone issuance of most contracts and grants."[36]
- The National Air and Space Administration "will increase basic research support from $685 million to $875 million this year . . . however, next year all grants for university research and facilities will be reduced."[37]
- "The Air Force places all research contracts on hold in the following departments: AEDC, AFSWC, BSD, AFWTR, AFFTC, AMD, ESD, SEG, AFMDC, APGC, RADC, RTD, AFETR, ASD, SSD, OAR."[38]
- Department of Defense "appropriations this year cuts $12.8 million from Research and Development funds and directs the Department to take this primarily from colleges and universities."[39]
- The Atomic Energy Commission "faces a pretty Goddamned big cut—conservative estimates range from $86 million to $114 million, much of it will probably come out of smaller and less established programs, such as nuclear medicine."[40]

Cutbacks in federal science policy hurt all of the sciences, but it was especially harmful for biological scientists at the three research universities in the Bay Area. Later that same year, the DOD and the NSF announced that "programs not [in] California, [which] receive almost one-third of the total research and development funds expended by the Federal government, will obtain most of the federal money available." The redirection of almost $300 million to "have-not" institutions departed radically from previous policy only a few years earlier because it promoted scientific equity rather than achievement and merit. These policy decisions, along with the popular rebellion against the scientific establishment, burned the biological sciences in the Bay Area for a time.[41]

Berkeley President Wellman tried to explain the cause of such perverse circumstances to his biochemistry department, confused still by the sudden turn of federal support: "Congressmen and Senators from the 'have-not' states see this state [California] as a clear target to shoot at."[42] Shoot at, indeed. Even representatives in California's state legislatures moved to consolidate and capitalize on newfound political momentum. Many, such as Democratic State Assemblyman Willie Brown from San Francisco and Republican State Senator Vernon Sturgeon, displayed a wicked genius for offering extravagant remedies for the public's pent-up frustrations with scientific research. The case of Jesse Unruh, a staunch Democrat from Southern California, is most instructive because he had, on more than one occasion, used his power as Speaker of the Assembly to direct enormous sums of money into the coffers of the University of California, much of which was used for scientific research. But in 1967, Unruh could only offer a searing rebuke of the entire system of higher education in California—to an audience of industrialists, no less, who normally supported the use of state funds for elite academic research—and compared the state's traditional support of the universities as similar to the way that Herr Dunderbeck treated his machine: "we have neglected the sausages."[43]

Unruh pitched his speech in terms of a democracy out of whack. "Who controls an institution of higher learning," demanded Unruh, "the administration, the faculty, the students, or a combination of all three? I am concerned when high officials—at all three levels—become incensed over the efforts of the Legislature to find the answers to such a basic question." Casting his net so wide that he condemned student protesters, university administrators, and academic researchers alike, Unruh decried the unwillingness of college campuses to respond to the will of the people: "How, then, the Legislature can help but be unavoidably involved in such an undertaking is beyond my understanding." Unruh stood defiantly behind popular support that ran "nearly 20–1 in favor of a harsh administrative and legislative crackdown on the universities." Advocating what must have been the worst possible nightmare that local researchers could have imagined, Unruh proposed "an almost complete re-allocation of resources." The biological sciences were certainly not the primary target of Unruh's attacks, but they were, in the minds of a reform-minded state legislature, part of the problem.[44]

Governor Ronald Reagan held the biological sciences in a chokehold too, slashing state money for medical care and education, which ironically united scientists and students in common alliance. Hostility toward Governor Reagan reached such intense levels at UCSF that medical school faculty and students mocked comic-tragically about doing research in the age of "AR"—"after-Reagan," as opposed to the glory years of "BR," or "before-Reagan." Reagan may have been a popular target for those who wanted more money for pure bioscience research, but in reality, both parties in the state legislature generally appreciated his aggressive stance and shared his distrust of the scientific establishment. In 1967, state representatives followed Reagan's lead and passed a bill that required every researcher "receiving state funds to report to the Department of Finance . . . so that the Legislature may be fully informed of the amount and impact of federal funds and state-supported programs." Scientists, who not so long ago enjoyed "reasonably wide latitude in determining procedures for compliance," found the new system of checks and balances to be a nightmare of epic proportions; the paperwork alone, complained one frustrated official, significantly "reduced the amount of time [he could] spend at the bench." District representatives also spoke less confidently in defense of pure research, and fewer citizens stood proudly behind the accomplishments of their homegrown heroes. And on some occasions, local representatives posed a greater threat to the traditional research establishment than the state or federal government. For instance, the Berkeley City Council decided to take the issue into its own hands and passed a restrictive ordinance that limited use of "potentially dangerous and pathogenic strains of bacteria." It was a dead law, because in their haste to handcuff runaway research, Berkeley City Council members did not bother to learn about or include in the law precisely which pathogenic bacteria they were referring to, or to differentiate which strain of *E. coli* was the source and solution for their angst. Across the bay, Palo Alto City Council members angrily opposed Stanford's offer to buy out Palo Alto's share of the hospital because, in the words of Shirley Temple-Black, the offer was a "giveaway," and gave Stanford faculty license to focus on research and ignore the medical needs of the local community.[45]

As bad as 1967 looked for local bioscientists, the future foretold an even worse fate. To begin, Richard Nixon appeared well positioned to capture the presidency. A calculating politician, Nixon held opinions about the biological sciences and basic research that mirrored those held by his predecessor and that overlapped with the psyche of his "silent majority": contempt for elites, ivory-tower academics, bloated budgets, taxes, inaccessible health care, government waste, the dole, and so on. As poorly as Nixon looked upon science, however, his Democratic opponents, such as Hubert Humphrey, spoke even more aggressively against basic research and promised to redirect more money into health care. Once in office, President Nixon extended the cuts to academic research initiated by the Johnson administration, the result of which drove many medical schools into the ground. For instance, by late 1969, 61 of the 103 medical schools in the country required federal assistance, twenty-nine of which had such severe financial need that their accreditation was in jeopardy. Some went so far as to speculate that opposition within the government went much deeper than the executive or legislative branches, but that peculiar personalities within Nixon's cabinet held personal contempt for their field, none more so, they believed, than Secretary of Health, Education, and Welfare Caspar Weinberger, who many in medical research dubbed "Cap the Knife" for his willingness to cut NIH budgets.[46]

A Season of Policy Reform

With heightened anticipation in this tumultuous climate, the Ninetieth Congress met. For all the alleged inscrutability of opinions on Capitol Hill, legislators had a general understanding of what their constituents wanted. The elderly and the unemployed wanted greater access to affordable health care. The middle-aged wanted greater protection against heart disease and cancers. Parents wanted to believe that children would not face genetic or inherited disorders. Many wondered what environmental pollutants had done to general health. And even without being fully aware of the consequence, the desire to increase the number of health-care providers meant greater professional and educational opportunities for a greater number of people. Certainly not all of the expectations were reasonable nor all of them destined to be achieved, but many voters shared a belief that the biological sciences could mend many of the wicked evils of modern life.[47]

Under such awesome expectations, Congress began its momentous task. Some of the bills under consideration did little more than cut spending across the board by a few percentage points. It was even easier to reject heavier state regulation. Nor could legislators allow scientists to keep control over the distribution or allocation of federal money. Committees also considered and then rejected "last dollar" financing—federal support individually tailored for the particular needs of each medical school or advanced bioscience department. They also considered and rejected providing indirect support to medical schools through the expansion of federal student financial-aid programs. Simply put, no one wanted a quasi-independent, bipartisan federal regulatory commission, especially if a temporary policy that reallocated funds fairly and purposefully could redress recent abuses.

The crucial solution appeared as a policy "trigger mechanism," what legislators called "capitation." According to capitation formula, Congress would establish a number of minimum requirements that a university must meet before its bioscience faculty could receive state funds. For instance, some of the more novel "pre-clearance" provisions were a requirement that at least 25 percent of a medical school's enrolled student body must qualify for nonresearch-oriented national health scholarship awards; fewer restrictions on foreign students entering medical professions; a 5 to 10 percent net increase in the number of enrolled medical students; fewer required research-based courses; and an increase in medical-training courses. The nature of capitation aimed to provide alternative funding mechanisms to medical schools so they could sustain basic research programs and at the same time respond to national needs; institutions that fully complied could use money to maintain basic research programs while those that met only a few would find their funds restricted. In sum, the frequency and degree to which medical schools met these capitation conditions would determine the amount and flexibility of their funding.[48]

A senior writer for *Science* caught wind of the coming changes and warned his readers that "there appears to be a good deal of emphasis on applied programs." Indeed, a steady legislative drumbeat began to pound out specific capitation requirements. Unlike years past, however, these bills offered funding with strings attached. Previous science policy allowed investigators to pursue any research topic, even if it had no immediate or obvious use. New science policies required clear statements of purpose. For instance, Senator Lister Hill, Chairman of the Appropriations Subcommittee for the Departments of Labor and Health, Education and Welfare submitted the Health Professions Education Assistance Act, which redirected appropriations away from unrestricted support of research toward "the immediate need for both construction [of new medical schools and programs], matching money and student loan matching funds." Congress also increased the budget of the NIH Research Institutes and Divisions by $139 million, but stipulated that $92 million must go directly to cancer research and $22 million to heart and lung research. Another Congressional committee told the NIH that it had a "moral commitment" to distribute Sloan-Kettering "block-grants" toward cancer research. Congress also made it clear that the NIH must increase the rate in which applied research projects received positive reviews; apparently, the standard 59 percent rejection rate was "a-less-than breathtaking batting average for Congress," admitted one NIH official.[49]

One of the more startling changes, and one that further illustrates the changes taking place in science policy during the late 1960s, was the new emphasis on applied research by the NSF—the science research agency that had been originally conceived and run to support classic basic research. By the mid-1960s, the once small, relatively obscure NSF had grown appreciably in size and scope, from an agency that dispensed small grants to individual-initiated basic research projects in the physical sciences into one that supported so-called "Big Science" programs in a variety of scientific fields, including molecular biology. In 1966, however, the House Subcommittee on Science, Research and Development was assigned oversight duties of the NSF to review its commitment to the basic sciences. The chair of the committee was Democrat Emilio Daddario of Connecticut, a Northeastern lawyer, decorated veteran of two wars, and tireless student of modern warfare from a stint in the Office of Strategic Services, where he analyzed German science and its contributions to war. Daddario entered Congress in the wake of *Sputnik,* developed an immediate appreciation for the sweeping impact science had on modern warfare, and had shown in years past a sincere desire to protect the foundation's basic research mission.

Yet, much had changed by 1968, including Daddario's view of basic research. He concluded, like many of his colleagues, that the NSF should begin directing "some research . . . in the national interest." Inspired by popular support, Daddario single-handedly introduced and capably guided through passage a bill that inserted a new engineering influence into this once pure research agency, making the NSF "more sensitive to the shifting winds of our national scientific climate and the government's role therein." With this amendment, basic research programs received smaller appropriations from the federal government while support for projects designated Research Applied to National Need (RANN) nearly doubled. The scientific community responded to the Daddario amendment as if it had radically altered the scientific landscape, which it did not; the NSF still allocated a preponderant share of its money to support for fundamental research. Yet, this slight shift in NSF policy meant that nothing in science was immune from the iron will of an imperious state. Practical, utilitarian, relevant, applied—however it might be defined, the desire to make the biological sciences more responsive was the leitmotif of new science policy. Thus, for the first time, bioscientists could apply to new divisions at the NSF, such as the Biological and Medical Science Division, and qualify for research support if they predicted unforeseeable application of their work; for instance, how protein synthesis might one day lead to a discovery that could "feed a teeming world" or lead to the development of new "pesticide production . . . for individual comfort and greater crop yields," as two early applications to the new "engineering" arm of the NSF proposed.[50]

In the eye of the legislative hurricane, however, few could match the California legislature's prodigious output. The number of bills related to health and medicine that were introduced and passed into law at the state level during 1967 has never been matched, before or since (Table 6.1). When state representatives finally adjourned in exhaustion, they could reflect upon a number of important pieces of legislation they had just passed. For instance, California legislators overwhelmingly passed the Casey Bill, which provided large state subsidies to assist the future fiscal needs of medical care programs, but "only by virtue of there being

TABLE 6.1. SELECTED BILLS SUBMITTED TO THE CALIFORNIA LEGISLATURE IN 1965 AND CONSIDERED IN 1966

Senate	
SB 17	Extends funding of research on cancer.
SB 407	Prohibits use of pesticides that can kill.
SB 495	Increases penalty for selling tobacco to minors.
SB 536	Prohibits placement of vending machines that sell tobacco products in a location used primarily by minors.
SB 543	Establishes a committee to study hospital construction and expansion program.
SB 608	Requires all transported food to abide by existing health laws.
SJR 19	Mandates an immediate study in fertility control.
Assembly	
AB 12	Requires all newborn children be subject to diagnostic tests for preventable disorders.
AB 16	Prohibits granting of licenses for the disposal of radioactive waste.
AB 22	Places the Dept. of Public Health in charge of state alcoholic rehabilitation program.
AB 85	Provides $25,000 for the establishment of an alcoholic rehabilitation program.
AB 138	Gives local health officers power to close areas deemed a menace to public health.
AB 219	Deletes as a felony and/or misdemeanor the distribution of birth control.
AB 258	Allows the Dept. of Public Health to "gratuitously" distribute prophylactics.
AB 259	Eliminates requirement for distribution of prenatal testing outcomes.
AB 260	Increases the number of members on the state hospital advisory board from seven to nine.
AB 261	Provides the Dept. of Public Health the power to extend or withdraw hospital licenses.
AB 349	Forbids sale of any product classified as poison to anyone under the age of 21.
AB 443	Requires pharmacist who sells *chloromycetin* to affix to the container a warning label.
AB 446	Requires prior approval of disposition of cremated remains.
AB 472	Establishes special programs for children who suffer from epilepsy.
AB 587	Excludes students who graduated from high school and who have not received a polio vaccine from being admitted to a California college.
AB 690	Requires establishment of safety standards for facilities used by the mentally retarded.
AB 691	Establishes regional centers to provide counseling for mentally retarded persons.
AB 747	Requires Board of Public Health to establish minimum standards for entrance into clinical laboratory technologists program.
AB 769	Establishes mental retardation program board.

AB 1300	Creates a division within the Department of Motor Vehicles to investigate motor vehicle accidents with fatalities.
AB 1305	Allows for humane or life-saving abortions performed by qualified physicians.
AB 1360	Requires persons new to the state to receive a polio vaccine before entering school.
AB 1448	Prohibits red or pink lighting over meat displayed for sale.
AB 1458	Permits corrections of errors in a certificate of birth, death, or marriage.
AB 1466	Requires minimum standards for any person who has received baccalaureate degree in clinical laboratory technology training.

Source: Robert Webster, Chief, Division of Administration, State Department of Public Health, Bureau of Health Education, "California's Health," 181–82, in BANC, CU-5, series 5, box 156:18, file: CA State Legislature, General, June 1964–June 1966. This incomplete list of health bills were all introduced during the 1966 Legislature Record.

patients . . . with an emphasis on Medicare and MediCal patients." Other pieces of legislation were simply improvisations, containing no trace of the sophisticated capitation machinery produced at the federal level. No matter; the intent was usually a symbolic one designed to convey a specific message to a stubborn audience.[51]

These first modest steps at a direct state role in guiding the practical application of bioscience knowledge represented an important and durable beginning. More aggressive legislation would continue to pour out over the course of the next few years, such as the Health Manpower Policy Alternatives in 1968, which was ostensibly a neutral process—revenues used to fund the HMPA still went through the NIH. Instead of distributing the money through their various basic research institutes, however, NIH officials established new programs that accommodated public concerns, such as Medical Education and Health Care (MEHC). In 1969, a near unanimous Congress adopted the Tyding's Family Planning and Population Act, which provided almost $1 billion for research on family, population, and service programs. In 1970, Congress overwhelmingly passed the Yarborough Bill, which authorized $50 million for 1971, $75 million for 1972, and $100 million for 1973 to "assist [medical schools] in establishing special departments and programs in family medicine and to promote the training of medical and other personnel in family medicine."[52]

The use of science policy to make the biological sciences more relevant was, to a great extent, contrived to meet a political—rather than a scientific—agenda. But since bioscientists chose to ignore popular opinion, politicians had few options left to them. Naturally, bioscientists responded to the shift in science policy by lambasting the susceptibility of politicians to the demagogic appeals of the masses, or the ignorance of these same masses for believing that science is malleable. Out of habit, they also felt as if they had been abandoned by physicians who did not adequately defend the importance of pure research. Indeed, a group of bioscientists at Stanford blamed the precipitous decline of the medical school's reputation on the "gross imbalance of talent" between research and health care—apparently, their judgment about the skill set of the two groups rested on an assumption that required no further elaboration. When a statewide bond measure to build a new "Fifth School for the Basic Medical Sciences" at UCSF lost in an electoral landslide, bioscientists blamed university physicians for their failure to campaign on their behalf rather than admit that perhaps California voters no longer wanted to use tax money to support pure research. Having no medical group to attack directly, Berkeley bioscientists nevertheless waged war over a .15 FTE appointment that had been transferred out of biochemistry and into immunology—a pitifully inconsequential academic matter considering the bigger issues at stake. Attacking anything or anyone affiliated with applied bioscience research may have brought clarity to the confusion that had beset the biosciences in the Bay Area, but the sad fact remained that it did nothing to solve the budget crisis that threatened to tear entire programs and departments asunder.[53]

Suspended between an old order and a new one whose shape had yet to be revealed, biological scientists and university administrators gradually began taking control of their budgets. The easiest was slashing student aid, which all three research universities did with rare bureaucratic ease. Weighty matters such as travel allowance grants and coffee funds almost always fell under the budgetary knife too. In 1967, the Stanford biological sciences department sold its floating marine biology research laboratory, the *Te Vega*, and Arthur Kornberg gave so many speeches at events that he likened the task to "the qualities for mobilizing for war," which in a perverse sense, it had become. Berkeley approved of molecular biologist Donald Glaser conducting experiments in a privately owned biological laboratory off campus, and then crammed graduate students into his old laboratory on campus. UCSF sold obsolete experimental equipment to local grade schools and then replaced the equipment with donations from private firms. Sometimes it was the smallest cuts that hurt the most, such as the popular science outreach program for inner-city school children sponsored by Berkeley's bacteriology department. Despite these actions, the cry for greater research support grew ever more insistent. So desperate had all three universities become that each began experimenting with what could be characterized as coercive fundraising techniques—and sometimes with outright fiat. These new solutions often meant upending a cherished tradition.[54]

In an unassuming second-floor office located outside the inner sanctum of Stanford University's Campus Loop, Niels Reimers, a patent attorney and former engineer, launched a pilot technology-licensing program that encouraged university faculty to exploit their inventions for commercial opportunity. No one spoke in favor of Stanford's Office of Technology Licensing (OTL), and Reimers met more than his share of criticism for running the program, but there was no shortage of interested participants either, in part because Reimers had restructured the university's standard royalty agreement by increasing the amount that the academic inventor could receive from zero to 33 percent.[55]

Berkeley Chancellor Clark Kerr had a little more in mind than using the marketplace as a solution to the University of California's financial woes when he summoned lobbyists to ratchet up their large and well-oiled political lobby. Prodded by Kerr, needing unlikely pliancy from state legislators, UC lobbyists devoted painstaking energy to gathering virtually every conceivable piece of information on all elected officials in California that might help them win their support of state science policy: date and place of birth; preferred nicknames; spouse's name, date of birth, and preferred nicknames; parents' names and date and place of birth; and, of course, opinions about science, medicine, and education—any connections they might have to the University of California. They also gathered the same data on incumbents who lost elections and might run again as well as potential candidates who had yet to enter the political arena.[56]

Across the bay, acting UC President Harry Wellman authorized a new Gifts and Endowments Office that would raise funds through a patient correspondence program—recognition letters, monthly newsletters, "Thank You!" and "How are you feeling?" mailings, tours, speakers bureaus—by encouraging donations from people treated successfully at the UCSF Medical Center. Faculty opposition was intense, but primarily for superficial reasons rather than for deeper meanings about fairness and objectivity in science, research, and medical care. "When will the campaign stop?" some worried, while others opposed all fundraising by public institutions like the University of California because it lightened the burden of responsibility normally borne by California taxpayers and set a dangerous precedent for public education. Only John Saunders, the lame-duck president of UCSF, whose support for human biology brought about his downfall with the epic coup five years earlier, spoke about the ethics of the Gifts and Endowments Office: "are we a 'for-profit' or 'not-for-profit' medical center?" he often asked, and wondered aloud about potential conflict of interest if patients could influence their medical treatment with donations.[57]

These three programs—the OTL at Stanford, Berkeley's lobby campaigns, and the UCSF G&E Office—although disruptive in terms of traditional academic patterns of finding money for scientific research, signaled a simple but momentous shift in perspective. By taking these small steps, Stanford, Berkeley, and UCSF were not just acting out of economic necessity; they were capitalizing on opportunity—and few fields offered more opportunities for earning money than the biological sciences. Of the three, however, it was Stanford's OTL that pushed ideas out of the ivory tower and began treating them as any business would treat intellectual property. It would be a fateful harbinger for university administrators and biological scientists who would discover market values like never before.

Chapter 7
Crossing the Threshold

Important scientific innovation rarely makes its way by gradually winning over and converting its opponents: it rarely happens that Saul becomes Paul. What does happen is that its opponents gradually die out and that the growing generation is familiarized with the idea from the beginning.

—Max Planck, 1936

By the end of the 1960s, the overriding issue in the biological sciences had become the practical application of pure knowledge. The popular wish for utility had become the preferred answer for bioscience research, draining fundamental discovery of its experimental significance, moral preeminence, and disciplinary authority. Here was an attitude about the biological sciences—to call it a philosophy would be too much—distinctly different from the attitude of the experimentalist, who stewed in anxiety about the pressure to change and lashed out at Congress and the country about the importance of preserving objectivity and autonomy, not to mention the integrity of the field. Although concern for relevant research did not resonate for all biological scientists in the same way, by the end of the decade, this much is clear: changes were significant. At least for a time, a new direction for the biological sciences could be glimpsed in shadowy form, from how research was organized to who actually conducted the experiments.

As one of the newcomers aptly put it, "this was just the opening act." From the organizational and demographic changes poured new discoveries, among them environmental controls, prophylactic vaccines, surgical technologies, medical procedures, immunization therapies, and so on. Out of this maelstrom of discoveries, however, the one that will probably weigh more heavily in medicine, agriculture, industry—in human history—is, simply, one: biotechnology.[1]

A New Direction

The biological sciences at Stanford, UCSF, and Berkeley had grown during the two decades after World War II from an inchoate set of programs into an academic force. Pure research served as the strategic center for most of this period of expansion: it defined laboratory organization, determined technological needs, provided individual investigators with a driving purpose, and attracted enormous federal support.

By the late 1960s, however, the field was at a crossroads. Suddenly, it seemed as if everyone in the Bay Area was "taking a new look at the remarkable laser to determine its usefulness," "integrat[ing] electronics into medical implants or biomechanical processes," or performing other "audacious experiments [that] promise decades of added life." In direct response to new science policies, Stanford began designing "meaningful programs . . . that would enable and encourage basic medical-science students to interact more with the clinical sciences." Along a similar calculus of financial concern, Berkeley developed new academic departments, such as the "Department of Interdisciplinary Biology," which "link[ed] the basic body of knowledge in the biosciences with contemporary life science topics." Likewise, UCSF administrators dismantled many of their "Organizational Research Units" and then restructured them as "Clinical Investigative Units" to coincide with the objectives of new science policy.[2]

But the revolution in the biological sciences owed to more than just federal money. Precisely how investigators responded to new opportunities made a difference too. For instance, bioscientists at Stanford broadened their experimental interests. In 1969, a HEW task force on prescription drugs reported that more than one-seventh of all hospital days were devoted to the care of patients whose doctors prescribed for them harmful combinations of drugs: "the hazards resulting from therapeutic drug use far exceeds other hospitalization hazards." So, in response to the HEW report, Stanley Cohen at the Stanford Medical Center collaborated with investigators in the biochemistry department to build a drug interaction database that would allow physicians to prescreen potentially harmful drug combinations. At about the same time, Stanford administrators cut one full year of basic bioscience study from its celebrated five-year medical training program. A medical school dean described the decision to commit more resources to patient care as "a necessary catch-up phase" that would allow the university medical center "to meet the needs of patients in the community and the region." Left unsaid, Congress had just passed the Health Manpower Policy Bill, which qualified Stanford for an additional $9.1-million capitation subsidy.[3]

Bioscientists at UCSF generally focused on cancer-specific research, made available through Richard Nixon's War on Cancer. Perhaps no other legislative measure proved more lasting or consequential in terms of experimental focus, at UCSF or elsewhere: Rudi Schmid and Holly Smith, deans in the Department of Medicine, implemented ambitious, broad-based clinical cancer-research programs, Martin Cline introduced large-scale chemotherapy studies, William Reeves organized a research unit to test the response of tumors to various chemical agents in vitro, Werner Rosenau and dozens of students, postdocs, and visiting scholars examined the role of lymphocytes in the immune process, and university administrators used federal funds to expand the medical center's Visible Tumor Clinic and Tumor Registry, establishing UCSF as one of the leading cancer data centers in the country. Their opportunism paid enormous dividends. In 1968, the budget for the only dedicated cancer-research program at UCSF—the Cancer Research Institute—totaled a measly $251. A few years later, however, UCSF had become the happy recipient of, among other awards, $296,256 for five consecutive years from the NCI's Clinical Cancer Research Center, a $70,877 Clinical Cancer Medical Training grant from the PHS, $82,410 from the NIH Western Cancer Chemotherapy Group, $54,637 from the NIH to buttress the Tumor Registry Training Program, and an NIH grant of $23,232 for the study of leukemia.[4]

These early efforts to redirect the biological sciences toward greater practical application furnish a startling demonstration of the willingness of some investigators and administrators, at least for the moment, to submit to the will of Congress and the people. But how deeply had this handful of new experimental programs affected the biological sciences in the Bay Area? Implementing a single experimental research project could in fact be achieved with a few simple and straightforward administrative procedures, none of which necessarily signaled a larger intention to radically embrace applied bioscience. Moreover, nothing prevented an individual investigator from reporting in their grant applications a deep commitment to practical application, and then redirecting the soft money awards toward their own projects and interests; a few even admitted years later that they may have "massaged" their stated experimental objectives to meet the expectations of grant-review committees. Altogether, the implementation of a few practical experimental programs did not permanently establish "human-centered scientific research projects," the "reestablishment of public health," or the drive to "meet the problem of disease," as many of the new federal grant applications requested.[5]

And yet, the core of these new experimental programs established a coherent base from which the biological sciences could then be reorganized at the department level. A close examination of three bioscience departments in particular—the biology department at Stanford, the biochemistry department at UCSF, and the Institute for Experimental Biology at Berkeley—shows how historical and political circumstances, and age-old competition between generations and between practitioners of basic and applied research, framed the biological sciences in the new era. Indeed, it is within the difficult context of personalities, academic politics, disciplinary instability, and new federal policies that efforts to reorganize the biological sciences must be understood.

Since World War II, Stanford administrators had tried, with little success, to persuade staff in the biology department to conduct "harder" bioscience research. To encourage interdisciplinary work with physicists and chemists, administrators such as Frederick Terman had promised financial support, inside connections to federal science agencies, and open access to physical science laboratories on campus. Stanford administrators could have scarcely been more explicit about their expectations.

But Stanford's biology department had come from a formal naturalist tradition, committed to the "traditional study of plants and animals for themselves alone." The intellectual gulf that separated these two sides—laboratory research and naturalist studies—prevented either from fully realizing an ideal program. In the late 1940s, Dean Terman asked microbiologist C. van Niel to emphasize more fundamental research in his marine biology program, but the proud investigator had consistently spurned offers of financial assistance, graduate student support, or staff. Following van Niel, Victor Twitty had run the biology program from 1954 until 1963, and much like his predecessor, he too refused Terman's appeal for a "harder kind" of fundamental research in biology; rather, he promoted undergraduate instruction among his faculty. In general, both van Niel and Twitty believed that Stanford administrators should judge the biology department not as a research center, but as an educational program, a consistent vision that left the two sides at a decades-long impasse.[6]

By the early 1960s, the pinnacle of popularity for pure bioscience research, the decision of the biology department to remain committed to the classical naturalist format posed a paradoxical choice for the department's younger, less established faculty. Associate and assistant professors such as Clifford Grobstein, Charles Yanofsky, and Donald Kennedy recognized that although the department already stood as a strong program, it offered little opportunity for less-established faculty to advance their careers through pure research. These younger faculty members considered zoology, botany, and heavy doses of genetic husbandry as increasingly anomalous, and furthermore, concentrating in such specialities would prevent them from entering burgeoning fields like molecular genetics. Within the scope of contemporary basic bioscience, therefore, the commitment of Stanford's biology department to classical naturalist studies offered diminishing professional opportunities. To use a simple analogy drawn from biology, younger faculty believed that they had to conduct pure research instead of naturalist studies, or their professional careers would die.[7]

The 1962–63 academic year was about to open, and still the general purpose and direction of research within the biology department remained hopelessly undefined. Desperate to rally faculty around more contemporary research questions, Stanford administration and faculty in biology asked their youngest colleague, Clifford Grobstein, to reorganize the department so that it would be well positioned for the future. He had enjoyed a precocious success, thanks in part to exceptional energy and ability, and in part to a gift for making himself agreeable to people on all sides of the research question. He had a doctorate in zoology from UCLA, which earned the respect of the traditionalists, and although he was not a medical doctor, he was tireless in his promotion of the Stanford Medical Center as a training ground for something he called the "physician-scientist." Most importantly, however, by cultivating a grand research vision, much like the university's foremost experimentalists Arthur Kornberg, Paul Berg, and Joshua Lederberg, Grobstein had become the leading figure within the new field of developmental biology—some would later call him the "founding father" of the field.[8]

Grobstein wanted to promote a balanced approach to biological research, but to modernize the department he preached strict laboratory experimentation. "There was nothing sacrosanct about naturalist studies," Grobstein often said. Naturally, older faculty challenged his proposal by appealing to tradition, charging inefficiencies, or gratuitously pointing out the corrupting influence on scientific research when research support came from federal policy. Old and new ideas about research were once again defined by the professional tendencies of two generations of investigators; senior men, who assumed that biology should remain the same as it was in the past, while junior faculty wanted to conduct more fundamental research. However, enfolded within the dialogue between the two sides was a development that had even greater consequence: by 1962, the department had grown younger and more rigid in its commitment to pure research. Now a majority, younger faculty thus controlled the direction their department would take.[9]

Grobstein and his younger colleagues struck boldly. They changed the name of the department from "biology" to "biological sciences," strengthened physical and chemical science requirements for both the undergraduate and graduate curriculum, accepted more graduate and postdoctoral students trained in laboratory research, and established a rigorous federal grant application program. Casual observers noted at the time that the new biological sciences department would "go unrecognized by its colleagues of just a few years ago." Indeed, some of the newer staff in the department conducted research that was difficult to distinguish from that of the mathematician, chemist, or physicist, and so impressed the purist Arthur Kornberg that he even offered certain members in the biological sciences department the opportunity to use the biochemistry department's "electron microscopy, X-ray crystallography and various newer optical methods in the study of biological macromolecules."[10]

But the perceived need to elevate fundamental research in a traditional biology program was not the real issue, as Grobstein and the rest of the department soon realized. Indeed, Grobstein's reorganization of the department occurred at virtually the same time that popular support for fundamental research had begun to recede. It did incalculable harm to the younger faculty who had committed to basic research that departmental reorganization had occurred at such a historically and unstable moment.

At this critical juncture, the momentum that had shifted toward pure research began to shift back again. "It is an undeniable fact," noted a reluctant Grobstein in late 1964, that "the scope of biology now ranges from viruses to society." "If science and the humanities constitute two cultures," offered Grobstein's colleague Donald Kennedy, "then biology stands with its foot in one and its head in the other." To their credit, junior faculty in Stanford's biology department recognized sooner than most that public and federal support of basic research was suddenly less certain in the mid-1960s than it had been throughout their careers. Perhaps fatefully, they could adopt this more moderate position because neither their professional careers nor the structural changes that they had implemented less than a year earlier had had time to congeal.[11]

A few weeks later, when winter quarter opened in 1965, the new biological sciences department announced their somersault toward a wider range of research questions, studying "the molecular origins of life to the role of intelligence in guiding human evolution," or "life dissected conceptually from populations down to molecules." The department's new core curriculum reflected these fused objectives. For instance, courses such as Human Biology anchored traditional "life-centered" questions to a molecular view of life and still took into account contemporary social studies concerns such as "human population problems."

Moreover, students studied traditional biology topics such as heredity, and then received extensive training in laboratory experimentation at the cellular level, where "harder" bioscience disciplines such as biochemistry, genetics, cell physiology, and embryology comingled. By connecting the life sciences with the theoretical and the scientific, Stanford's biology faculty encouraged larger, more complex macromolecular research programs—a heretofore ambitious leap into what one observer described as "an untapped field, pregnant with relevance and possibilities." Not everyone supported this new curriculum. Those committed to pure bioscience research, such as Arthur Kornberg in biochemistry, scoffed at the biology department's ambiguous experimental objectives when there was so much to learn from simple bacterial, procaryotic, or phage studies. Kornberg may have been reasonable, but Grobstein's new department paid enormous dividends; before the decade was out, the department of biological sciences at Stanford University received, among other awards, $1,800,250 from the NSF to build a new center for research, one of the largest grants ever made to support biology.[12]

The academic coup that ousted John B. deC. M. Saunders from his tenure as dean of the UCSF Medical School in 1964 also finished many other academic careers, including that of David Greenberg, the longtime chairman of the biochemistry department. The faculty that remained may have wanted to remove all obstacles that restricted pure research, but it was not Saunders or Greenberg that created the obstacles in the first place. The UCSF Medical School at the time of the coup was still in many ways a ramshackle, disarticulated assemblage of factions—a small dedicated group of pure researchers unable and unwilling to calm their feud with a much larger number of physicians. The faculty in the biosciences had insufficient numbers to dictate university policy, so they relied on unrelenting determination to shape their own departments according to their highest ideals.[13]

It was only logical, therefore, that the search for a biochemistry chairman should become the staging ground for conflict. Because of the peculiar significance of biochemistry in the biosciences, and because of the peculiarities of academic search committees, the faculty in charge of selection came from the most fanatical base of purists, and they chafed at any suggestion of compromise—the future of UCSF, the biosciences, and their professional careers depended entirely upon the protection and expansion of pure research, or so they thought. Not surprisingly, all this passion made it nearly impossible to find a top-flight biochemistry candidate who had the leadership skills to drive the university through the cloudy bioscience future. But rather than rationally solve internal problems, the search committee chose the path of less resistance: they would attract the perfect leader by simply outbidding all other competitors. Of course, they left nothing to chance: salary, new staff hires, and additional laboratory space were all important issues at an urban university such as UCSF.

As designed, the recruiting package attracted the attention of many capable and interested candidates, but not enough to overcome the confusion that had beset the UCSF Medical Center in the first place. Between 1964 and 1966, seven separate candidates received an offer to chair the biochemistry department, and all seven refused. A year later, the search committee promised 12,000 square feet of additional laboratory space and three additional staff, but no one would accept the chairmanship, so they increased the offer again, this time to 14,000 square feet and another faculty appointment, and still no one came. By 1968, the offer had surged to 29,600 square feet, up to fifteen new appointments, to no avail. In all, four committees led by five different personnel worked for nearly eight years to find one chairman of biochemistry, and all they had to show for their maximum effort were nine formal rejections and countless more who did not bother to apply.[14]

The plight of the search hurt deeply, and yet, a silver lining could be seen in the wave of rejection letters. Gradually, certain faculty members, once rigidly committed to basic research, began to question their own ideals about the supremacy of fundamental knowledge. Julius Krevans, who came out of the era of pure research, noted the "curricular convulsions" in the new bioscience climate, and saw in the counter culture and policy realignments a possible solution: "There [is] hardly a month that goes by without some critical editorial appearing on the failure of medical schools to address such topics as the principle problem of nutrition, infectious disease, genetic disorders." There have been a number of impressive fundamental discoveries in recent years, Krevans said with much pride. But, he also conceded, "there has been precious little progress in academic medical centers in our understanding of some of the most fundamental problems, such as nutrition, alcoholism, and the inter-relationship between internal metabolic processes concerned with nutrition and the external effects of deviations from normal nutrition." To the dismay of his colleagues, Krevans suggested that perhaps "it would make sense to make practical research questions, such as infectious disease, a major charge for [the biochemistry chairmanship]."[15]

Krevans's suggestion that the search committee relax its rigid commitment to pure research constituted a historical pivot for UCSF and was not necessarily welcomed. Traditionalists such as Ernest Jawetz, chair of the microbiology department, felt "disheartened and demoralized" by Krevans's proposal, and he threatened to resign from the search committee if the clinical sciences ever became a primary consideration. Some did in fact resign, leaving with the hope that "the basic sciences could someday be reintroduced during the so-called 'clinical-years.'" Manuel Morales, perhaps the most passionate purist on the faculty, declared that anyone who conducted research in the clinical sciences had committed an "intellectual sin."[16]

But by 1968, an emergent band of new and younger faculty had developed an intellectual alliance with physicians and the clinical sciences. Microbiologists Leon Levintow and Mike Bishop had become frustrated with UCSF's "sterile research environment," and Herbert Boyer, a promising microbiologist just out of the University of Pennsylvania, threatened to leave UCSF because, among other reasons, evaluations for tenure placed too much emphasis on fundamental discovery. Ironically, the endless debates and constant failed searches blurred the ideological line that divided basic and applied research, as was also happening in the new biological sciences department at Stanford University.[17]

When Krevans challenged the search committee to broaden their ideal of what constituted good and proper bioscience research, they found that Bill Rutter, a biochemistry professor at the University of Washington, suddenly qualified as an exceptional candidate. Rutter's professional dream was also the popular dream of applying pure knowledge. His conception of bioscience research was married to his ambitious, restless temperament—a temperament that saw great promise in the biosciences. Among the most vivid evidences of Rutter's ambitions were the practical research questions he consistently sought out, even while fundamental research reigned supreme. For instance, as an undergraduate in the 1950s, Rutter double-majored in social science and biology, studied medicine at Harvard, and did his doctoral work in biochemistry at the University of Illinois on a most un-pure research topic, galactosemia—a genetic disease in which the body fails to metabolize galactose sugar. As a postdoctoral student at Wisconsin, Rutter conducted a number of experiments in which he applied bacterial genetics and enzymology to various agricultural problems that had plagued the state's dairy industry. He also spent a little more than two and a half years as a postdoc at Stanford, not coincidentally under the direction of Clifford Grobstein in the new biological sciences department.[18]

It is plain that not everyone in the biosciences at UCSF was impressed with the scientific rigor of Rutter's research, but whatever other frustrations they may have had, his selection as chairman did have organizational rationale. Indeed, once in place, Bill Rutter showed remarkable scientific talents, but more important, he had greater skills as a team builder. He encouraged his faculty to study eukaryote biology and pursue complex research topics: the structures of chromosomes and the regulation of gene expression, the mechanism of hormone action, antigen induction, and perhaps more significantly, "since all human processes are directly or indirectly governed by genetic mechanisms . . . there is no practice of medicine which can be fully appreciated . . . without a sound knowledge of . . . molecular genetics." New, young, talented investigators such as Harvey Eisen, Howard Goodman, Jim Spudich, Reg Kelly, and John Watson joined Rutter's team. As perhaps the highest compliment, Gordon Tompkins, who had been offered—and had turned down—the same chairman position that Rutter now occupied, gladly accepted an offer to join the UCSF faculty. Many of these new hires selected by Rutter had joint Ph.D./M.D. degrees—a natural interface between two factions, and a sharp contrast with the purity of bioscience education and training in the 1950s and early 1960s. The arrival of Rutter also helped convince microbiologists such as Boyer, Levintow, and Bishop to stay. The merging of basic and clinical research, which quietly began as a reluctant compromise, quickly became the standard for research at UCSF and would one day serve as the leading edge of a program called "clinical scholarship."[19]

The decision to appoint Bill Rutter as chairman of the biochemistry department was also a decision to blur the boundary that had long separated pure from applied research at UCSF. That simple but momentous shift in perspective was the newest thing of all at UCSF, and it brought a much firmer financial footing too. While the NIH was curtailing many of its generous training grants, it gave nearly $775,000 to promote UCSF's new "clinical-scholars" program; other federal agencies granted additional capitation funding because the medical school increased the size of the incoming class by exactly 5 percent—from 128 to 135; despite an enormous budget deficit, the California state legislature gave the UCSF Department of Microbiology a one-time grant of $1,879,100 to explore "complex human processes," an award that the federal government subsequently matched with $2,816,300. It is hardly surprising, therefore, that the UCSF faculty and graduate divisions in the biological sciences began to race up the rankings, which attracted an even higher caliber of research faculty who would soon win, in a relatively short period of time, three Nobel Prizes.[20]

In 1963, Choh Hao Li, chairman and lone tenured faculty member in the Institute of Experimental Biology at Berkeley, announced that he had isolated and purified his sixth pituitary hormone, lipotropin. The magnitude of such a feat is clear considering that only one other person had ever purified a hormone, and that person was not coincidentally a student of Li's. The purification of lipotropin should have been a reason to celebrate; however, Li's colleagues at Berkeley acknowledged but did not rejoice in his success. As they perceived it, endocrinology was a scientific field that came out of the clinical sciences, which meant that Li's research was completely unsound, and they put enormous pressure on him to change his scientific topic. When that did not work, Wendell Stanley tried to "promote [Li] out of the Virus Laboratory," then later University Chancellor Clark Kerr threatened to discontinue the Institute for Experimental Biology because it did not fit with Berkeley's commitment to pure research. Things got infinitely worse for Li, of course, because he became perceived as less qualified with each professional achievement.[21]

The distorted values of Berkeley's bioscientists reached its peak in July 1964 when Li and two of his research assistants submitted an article about his work with lipotropin to the *Journal of Biological Chemistry*. Although the details surrounding this article remain obscure and may never be known, it is clear that Li and his colleagues failed to reference in their final draft the results of an experiment conducted in Sweden that was under review and scheduled for publication in another journal. Perhaps it was professional sloppiness, but Li's students believed then and now that his error was due to a language barrier—apparently, Li still had a difficult time reading scientific literature in English. Regardless, his failure to properly cite an experiment in progress exposed him to further attack. Scientists from all over chastised Li for lack of professionalism, but the harshest criticism came out of Berkeley. A colleague in biochemistry admonished Li to "behave [him]self," while another wrote a scathing letter that "LIPOTROPIN" was probably a "protected trademark." One went so far as to call hormone research a "stupid and childish intrigue." Clark Kerr, sensing the opportunity to finally remove Li and bring an end to a soft research topic like endocrinology at Berkeley, quietly "resettled" the maverick scientist at UCSF, where endocrinology fit into the clinical sciences more appropriately.[22]

C. H. Li's travails at Berkeley are only half the story. In 1969, five years after transferring from Berkeley to UCSF, Li and his laboratory assistants assembled a highly complex synthetic version of human growth hormone (HGH) that was biologically active and could promote the growth of bones and muscle tissue. Rather than ignore or criticize the work, however, journalists waxed eloquently about Li's creation of HGH. One described it as no less than a panacea for most of the world's problems. Others clearly saw specific applications: "it might now be . . . possible to tailor-make hormones that can inhibit breast cancer." Li's discovery of synthetic HGH "constituted a truly . . . great research breakthrough [that had] obvious applications," ranging from "human growth and development to . . . treatment of cancer and coronary artery disease." Desperate letters poured in too; athletes wanted to know if HGH would help them become faster, bigger, stronger, and dwarfs from all over the world begged for samples of HGH or to volunteer as experimental subjects. Unlike at Berkeley, Li's discovery made him a hero at UCSF. None other than UCSF Chancellor Phillip Lee described Li's discovery as "meticulous, painstaking, and brilliant research" and then tried to capitalize on the moment by asking the public and their political representatives to increase federal support of bioscience research. "Research money is dwindling fast," repeated Lee to anyone who cared to listen. "We've proved that synthesis can be done, now all we need is the money and time to prove its tremendous value." It is not surprising that federal and state money began to pour into Li's lab. What is shocking, however, is how quickly Li achieved scientific acclaim, not because he changed, but because the rest of the world around him changed so much.[23]

The transformation of the biology department at Stanford, the biochemistry department at UCSF, and the Institute of Experimental Biology at Berkeley and UCSF are examples of institutional flexibility: each department came from a traditional scientific starting point in naturalist or clinical sciences; each department forced its way greenly up between the cracks of pure bioscience research through to the mid-1960s; each department recognized the popular and policy trends and then made room for applied research; and then each department showed an uncanny ability to batten pure and applied research into a unified research program.

But something more crucial than departmental organization sits at the center of this bioscience revolution: the arrival of a new generation of bioscientists made the shift toward applied bioscience research more durable. This final piece of the puzzle must be considered alongside any new programmatic or departmental structure. Relevant research in the biosciences took hold because it suited the concerns of so many young investigators.

A New Generation of Bioscientists

In the 1960s, a new generation of bioscientists began entering university laboratories in the Bay Area—new because they were concerned with far more than acceptable laboratory practices, scientific methods, or the search for bioscience truths. How should David Gelfand balance his activism in the civil rights movement with his research on restriction enzymes? Should Janet Mertz allow her environmental concerns to affect her recombinant DNA work with *E. coli*—should she worry that she might inadvertently produce an uncontrollable virus? Should—or could—"Wild" Bill Holmes consciously separate his communitarian lifestyle and political radicalism from his research in molecular biology? Did Frank Lee's colleagues know that he was born and raised in Communist China—more important, did they care? Why did Robert Helling first begin "thinking about some way to start manipulating genes" when he left the Midwest and arrived as a postdoc at UCSF—and lived among the hippies in the Haight-Ashbury district of San Francisco? Should Peter Lobban expect a promotion even though many of his laboratory superiors at Stanford did not necessarily support his pathbreaking genetic engineering experiments? Should Richard Mulligan hide his past affiliations with the Communist Party in order to work in Paul Berg's laboratory? Did the LSD that Kary Mullis ingest while a graduate student in Berkeley's biochemistry program advance or slow his Nobel Prize–winning work with the polymerase enzyme—ironically, the same enzyme that helped Kornberg win his own award a decade earlier? Should Osamu Hayaishi, his wife, and children return to their home in Japan, or should they accept the uncertainty of another two-year postdoc assignment? Should Mary Betlach behave any differently as the only woman in a predominately male lab?[24]

The availability of federal money may have turned the Bay Area into a bioscience research center, but success did not mean that the new generation of investigators had it easy. In innumerable ways every day in the labs, senior and assistant researchers, technical assistants, and administrators confronted choices that exposed their deepest concerns and loyalties. Experimental programs and departments looked very different from just a few years earlier, but to truly understand the lasting impact of the counter culture and new science policy, it is important to do more than measure the magnitude of departmental change. It is necessary to ask whether the people in the biosciences were any different during this revolutionary period than they were a generation earlier.

In the late 1960s, the older generation of investigators who were committed to basic research confronted a less traditional scientist led, in the main, by their own students. The natural disequilibria caused by dramatic scientific discoveries, popular dissatisfactions, and the tightening of financial support had certainly destabilized the biosciences in the Bay Area, but the arrival of this new generation of researchers also caused great disruption. Their arrival, and their decision to apply their craft toward something more than fundamental discovery, bore significance for the future direction of the field. Of all the changes taking place in Bay Area bioscience laboratories during the late 1960s—programmatic, institutional, and even cultural, political, or economic—the arrival of this new generation of investigators wrought perhaps the most real and lasting change of all. This was a new breed of researchers entering Bay

Area bioscience laboratories in the late 1960s, and they clashed, sometimes mightily, with their predecessors.

To review, in the years immediately following World War II, the bioscience community in the Bay Area was a small and talented group driven by an incessant desire for autonomy—characterized, in their mind, as the freedom and opportunity to direct all their professional energy toward bioscience research that emphasized the search for fundamental principles. The entrance of physical scientists into the field and the effect that technological change had on their research—usually a slow and inconspicuous process—was rapid and immediate. As we have seen, the explosion of personnel and technology produced a hurried accumulation of bioscience knowledge and, in turn, an even greater emphasis on fundamental discoveries and the need for more resources to conduct that research.

What is also striking about this early era of pure research is the homogeneity of the people involved, especially at the elite levels. In general, almost all of the top-level investigators in the Bay Area grew up on the East Coast, got married while in graduate school, and earned their Ph.D. during the interwar years, in particular between 1929 and 1942. The intellectual and technological contributions of the physical scientists also created a rigid professional filter that required this generation to understand physics and chemistry. Only a handful of investigators in the Bay Area had an M.D. degree—the two at Stanford were trained in European settings that encouraged more research-oriented medicine and five of the six bioscientists at UCSF trained in the United States Army Specialized Training Program which accelerated the production of physicians during World War II but did not require or emphasize patient care. Those born outside of the United States overwhelmingly came from northwestern Europe, a byproduct of the McCarran immigration act, which limited ethnic diversity in the laboratories as much as local social prejudices and the desire to weed out Communist sympathizers.

Politically, this older group considered themselves loyal Democrats, but that characterization also needs qualification. On the one hand, they revered the memory of FDR and the New Deal and appreciated federal support for their work, but they also preferred smaller and less intrusive government. Beneath their liberal veneer—contemporary observers might call them neoconservatives—lay a politics of self-interest. They would eventually oppose the Vietnam War, but their humanistic rhetoric could not hide their deeper sense that the war was wrong because it sucked up funds they believed would have been better spent on their own basic research programs. They also proved quick to vote for a Republican if the candidate seemed sympathetic toward—or less hostile to—their work. For instance, most of Stanford's biochemistry department were registered Democrats but supported any candidate who in turn supported their own work, regardless of their political affiliation; many actively campaigned for the Republican Paul McCloskey and opposed the Democrat Shirley Temple-Black because "the little princess" wanted the Stanford Medical Center to commit more resources to patient care and less to scientific research; some also supported Richard Nixon because Hubert Humphrey promised to cut federal funding for basic research. Moreover, the elite investigators at Stanford, Berkeley, and UCSF may have been sympathetic to the civil abuses inflicted upon blacks and other nonwhite minorities, but they also actively opposed affirmative action–type minority enrollment programs.[25]

Bay Area bioscience laboratories in the early post–World War II era also had a significant number of second-generation immigrant Jews who grew up in or around Greenwich Village in New York City during the 1920s. Heirs of a socialist tradition imported from Europe, they appreciated federal support but were conditioned to distrust authority by the recent historical memory of persecution and had been brought up with a healthy respect for serious ideas, which informed their desire to prevent intrusion by the federal government into their laboratories. Anti-Semitism goes far to explain how or why so many investigators in fundamental research were Jewish; many in fact recount exclusion from medical schools on the West Coast, in the Midwest, and at its worst on the East Coast at institutions such as Johns Hopkins. Apparently, during the interwar years, medical schools found it more tolerable to allow a Jewish student into a research laboratory than to practice medicine on a patient in a hospital.[26]

Within this exclusive group existed another internal boundary: except in very unusual cases, the community was predominantly male. The world of the biological sciences may have seemed spacious and full of possibilities, but in the main, women had difficulty establishing themselves within the basic research community. A few notable exceptions existed, but even these illustrate the barriers that prevented women from full participation. For instance, Agnes Fay Morgan, one of the world's most prominent "vitamin hunters" of this era, was trained as a chemist at the University of Chicago but could obtain an appointment only at the junior level in Berkeley's home economics department, hardly an outpost for advanced bioscience thinking. Other women who conducted fundamental research, such as Miriam Simpson and Marjorie Nelson, found themselves "phased out" or transferred into clinical programs at UCSF or Stanford, where it was considered appropriate for a female scientist to pursue practical research questions. Rosalind Franklin, considered by many the best x-ray crystallographer in the world for taking the first "photographs" of DNA's double helix, obtained a short-term postdoc appointment in Wendell Stanley's BVL; however, she applied to Berkeley because her mentors James Watson and Maurice Wilkins forced her out of "their" laboratory in part because they were uncomfortable with her "unacceptably aggressive"—or masculine—behavior. Stanford appointed more women into the biosciences than Berkeley or UCSF, but all three women at Stanford were married to tenured faculty in the department: the Kornbergs, Lederbergs, and Herzenbergs. If women entered the basic bioscience laboratory, they did so most often as secretaries, or at best, temporary lecturers or technicians.[27]

How exactly women were excluded from more prestigious basic bioscience laboratories during the early post–World War II era is unclear too; it is difficult at this time to determine whether women were rejected for their gender, their inexperience, or because they did not bother to apply because they either anticipated rejection or simply found the rituals of basic research uninteresting or irrelevant. In any case, the intersection of women and applied research became self-perpetuating: pure research was privileged over applied, as were men over women; men conducted "harder" fundamental research projects, while the experiments conducted by women were considered "marginal"; the body lay at the heart of applied research, so naturally men should conduct pure research while women could engage the body in applied research. The most socially progressive defense of women in the biosciences grew from this set of assumptions: women not only belonged in practical or applied bioscience fields such as bacteriology, immunology, or metabolism, but they might even be better suited for it than men, or so said Wendell Stanley, Arthur Kornberg, and countless more.[28]

As has already been noted, by the late 1960s, the entire system—a product of disciplinary competition, public opinion, federal policy, and of course, revolutionary bioscience discoveries—had become destabilized. But there is also much truth to the historical cliché about the "alienation" of young investigators from their elders, their goals, and values. To be sure, much discontent between the two generations of investigators stemmed from the inherent tension that occurs when two people desire both autonomy and collegial control. Professional autonomy may have meant something entirely different to the new generation than the search for esoteric knowledge, driving professional elitism may have grated against the new generation's egalitarian assumptions, and "affluence" in bioscience laboratory may have become an economic or psychological force too overwhelming to tolerate. Perhaps it was simply a combination of explosive forces—more competition, less money divided among a wealthy population, new ideas, driving insecurities and so on—that antagonized the subversive or renegade researcher.

Certainly the two generations had some things in common. For instance, both generations of bioscientists in the Bay Area universities came from educated segments of the middle class: largely urban and suburban and financially comfortable, a significant number also had a father engaged in a professional career. Many were also Jewish *or* gentile, born in the United States, and went to urban or large public universities. Social relations between senior and junior faculty remained virtually unchanged too: the younger generation recognized that their professional fortunes depended on the success of the laboratory; the desire for professional success made everyone respectful of scientific authority.

While there were a few similarities between the two generations of investigators, the new generation of bioscientists nevertheless constructed identities that in many ways contrasted with the identities valued by the previous generation. For instance, the new generation of investigators formed guild-like professional organizations, such as the Salt and Water Club, or BANG (Bay Area Neurophysiology Group), much like Steve Jobs or Steve Wozniak formed the "Homebrew Computer Club" before launching Apple Computers or the "People's Computer Company." The training that both generations received was different too: bioscientists in the 1950s typically had advanced training in physics or chemistry while those entering the biosciences in the late 1960s often came from an engineering background or by way of medicine—fields that are inherently pragmatic and practical. By no means did this new generation challenge all of their predecessors' curious laboratory social patterns, complex institutional boundaries, or passionate commitment to fundamental research. But they did find it reasonable to expect that professional responsibility went beyond the laboratory; that assumption alone threw into the open questions about virtually every other facet of this tight-knit community. If nothing else, they accommodated the disruptive political culture and tightening political economy, either by choice or out of necessity. Whatever the core narrative, the mood to conduct bioscience research had shifted within Bay Area university laboratories, spearheaded by younger bioscience researchers.[29]

Another distinguishing trait of this new generation is the large number of investigators who were born in or trained someplace other than the United States or northwestern Europe. Bioscience refugees came in waves that swelled in size after the Immigration and Naturalization Act of 1965 abolished the old quota system based on national origin. At the same time, Congress relaxed loyalty requirements and established "special categories" for young students who had scientific or technical training, which made it even easier to obtain "favorable consideration" for resettlement. Many of these student immigrants came from countries like India and China that struggled with malnutrition and disease and oriented their research interests accordingly. In contrast to more typical migration patterns that both pushed and pulled new peoples to America, foreign-born investigators flooded Bay Area bioscience laboratories primarily because they found incredible professional and financial support for research that their native governments could never provide. This demographic trend had, quite possibly, a tremendous impact on bioscience research: the arriving foreign-born students inclined to address practical problems could now drink from the plentiful federal spigots that emphasized the practical application of bioscience research. It is possible that non-Western newcomers to Bay Area bioscience laboratories in the late 1960s placed greater value on practical application of fundamental knowledge and had not internalized the basic-over-applied hierarchy as their counterparts born in the United States or Western Europe had done.[30]

However, deep in the substratum of the bioscience upheaval remained one grim consistency: women had less access to research laboratories and held few positions of authority. Gender discrimination showed itself in countless ways in Bay Area bioscience laboratories during the 1960s: the number of doctorates going to women remained low, discriminatory admissions policies persisted, male graduate students enjoyed far greater upward mobility than their female counterparts, and the number of departments that had no women remained remarkably high. The continued stratification of women within the biosciences is all the more curious when one considers the surge of practical bioscience research in the late 1960s. Indeed, women should have found greater opportunities in the biosciences when applied concerns were on the rise; however, evidence drawn from Bay Area bioscience programs suggests that no effort was made to mobilize women for their practical skills or concerns at the highest levels of research. For instance, Stanford had no female tenured professors in the biosciences; Miriam Simpson and Linda Goodman were special instructors in biochemistry and microbiology at UCSF, but without Ph.D.s in the field, neither could earn a full-time appointment; and Ellen Daniel became the first woman appointed in the molecular biology department at Berkeley, though she was denied tenure and eventually left the program.[31]

Why qualified women had such a difficult time gaining entry into Bay Area university bioscience programs is hard to pinpoint. Certainly the manner in which male faculty evaluated the performance of women as investigators sidestepped real issues. For instance, one male faculty member at Berkeley often told single women in his program that they lacked skills or adequate training; at the same time he chastised married women for having a deeper commitment to their husband's career than their own. Having characterized single women as undertrained and married women as immobile, it is certainly no surprise that a woman in the biosciences at Berkeley might feel compromised: "I never felt that I was denied an opportunity because I was a woman, but I did feel that a lot of my ignorance about how to succeed came from being a woman, being un-mentored." Perhaps behind subtle ridicule, male bioscientists also took a cautious and compromising approach to gender discrimination.[32]

Perhaps also the strategies that women used to gain access worked at crosspurposes, such as the female graduate student who used the comfortable stereotypes of the day to justify her presence in the field: "Men will find [female bioscientists] feminine, not aggressive, and easy to get along with. . . . A woman . . . has unique qualities of warmth and understanding—qualities that can only be explained as a 'feminine touch.'" Such an anemic defense may have allowed a handful of women to gain entry into bioscience research laboratories, but it also played into powerful social pressures that funneled women into health-care professions rather than research; in 1967, women constituted 99.5 percent of the total number of students enrolled in nursing programs at both Stanford and UCSF, while approximately 15 percent of the students in Stanford's biochemistry program were women, 9 percent at UCSF, and there were no female graduate students in the biochemistry program at Berkeley.[33]

Barred from authority, women in the biosciences at Stanford, UCSF, and Berkeley often had to concentrate their efforts toward achieving limited gains. For instance, at the 1969 Association for Women in Science (AWIS) annual meeting in San Francisco, Judith Pool, a nontenured clinical researcher at Stanford's medical school, proposed that members of AWIS put all of their collective energy into increasing the number of women in science, but her motion was summarily defeated by her peers who favored a platform that ignored questions of access so they could focus on achieving equal pay for male and female investigators and administrators. At the height of the space program, Inka O'Hanrahan, a clinical biochemist at UCSF, established a subchapter of NOW and focused all of the organization's attention on getting women admitted into what she considered the more prestigious aeronautical programs.[34]

The obstacles that women faced in Bay Area bioscience laboratories were certainly formidable, and yet, on a relative scale that takes into consideration the conditions that they confronted on a daily basis, they achieved some real significant advances too. Unarguably, women in the late 1960s began challenging the exclusive boundaries that had long set them apart. Though most women still worked in administration or as assistants—secretarial, custodial, nursing, and so on—they had far less tolerance for separation within the profession. They condemned, with a vigor rivaling general student protest against pure research, the serious lack of female medical students, men's "invisible privileges," and demeaning and offensive behavior. For instance, a group of female graduate students at Stanford denounced the unspoken pressure to "put-out-or-perish" that had plagued academic laboratories in the past, while female students at UCSF demanded that the editor of the student newspaper remove the series of work-study advertisements that listed "big-knockers" as a necessary qualification for a technical assistant position. More significantly, from 1960 to 1969, the number of women who enrolled as graduate students in bioscience programs rose by 96 percent.[35]

Despite lingering discrimination and double standards, the mere presence of women in graduate bioscience programs served as an unorganized albeit powerful counterforce that checked unrelenting gendered assumptions about the appropriateness of women in Bay Area bioscience programs. Moreover, the gradual increase in the number of women in graduate bioscience programs probably made it easier for Congress and President Nixon to pass the EEO Act of 1972, which effectively ended the Title VII provision that had formerly exempted all educational institutions from equal employment opportunity laws, and replaced it with Title IX, which extended the Equal Pay Act to higher education and banned sexual discrimination in any program of an institution receiving federal funding. Perhaps there is also a relationship between the growth of women in the field and the continued emphasis on practical bioscience research in the next decade.[36]

A New Experimental Direction and a Moral Dilemma

In the transformation of the biosciences, one of the great underlying shifts of view was the development of a new appreciation for practical experimentation. In the decades before, investigators spoke of pure discovery. In a sense, they had to, for many of the phenomena they dealt with—DNA, proteins, enzymes—had never been seen and rarely studied. Yet pure research had a deeper significance; at the same time that bioscientists sought pure knowledge, they used it to define their work in relation to impure practical research. As has been made quite clear, pure research had both purpose and meaning.

Then, beginning in the mid-1960s, the field began to move in a new direction, with new experimental programs, departments, and disciplines. The objective was, simply, to create experimental space for the practical application of pure knowledge. Applied bioscience had become an organizational and demographic fact, and the organizational and demographic bulge aided the applied state of mind. Practitioners contrasted the utility of their work to the irrelevance and destructiveness of pure research, whose practitioners had greedily manipulated the field, suppressed its benefits, and perverted the discipline. Applied bioscience research was an experimental focus of both purpose and reform.

Of all the applied bioscience research projects under way in this new era, however, one in particular stood out, both for its scientific novelty and its impact on humanity: genetic engineering. It was as distinctive for the experimentalist as it was as a scientific experiment. Social relations were adapted to realize the potential for abundant and significant experimental outcomes. Identity, livelihood, professional success—everyone was bound together around the experiments. The rules of behavior that had governed laboratories like the BVL—exclusivity, hierarchy, focus—had given way to a new generation of genetic engineers who proved especially open, ambitious, and impatient, a combination of purpose and personality that held within its grasp tremendous potential—and no little danger.[37]

At some point in late 1966, Paul Berg, biochemist at Stanford, made the difficult decision to shift his experimental attention away from research of simple, single-cell bacterial systems and focus on genetic expression and regulation of mammalian cells—at the very least, to study genetic diseases in humans. Against the backdrop of popular, political, and professional changes taking place at this time, Berg's experimental ambitions were not unreasonable. With hindsight it is possible to see that a few bioscientists considered genetic manipulation a real possibility too. For instance, as early as 1958, one Nobel Prize winner predicted that bioscientists in another era would "produce better organisms" through "biological engineering."[38]

Cautiously, Berg approached his mentor and department chairman, Arthur Kornberg, and mentioned his new interests. Kornberg lambasted Berg's proposal: "you're wasting your talent," accused Kornberg, "you're destroying your career." In a sense, he had a point too. The human genome is much larger and more complex than the bacterial genome, which made Berg's proposal to trace the expression of mammalian genes, not to mention its control, extremely complicated. In Kornberg's mind, the problem that Berg would inevitably confront was not "where should a person begin such an investigation" of genetic disease in humans, but how? Dramatically revealing his own estimate of the gravity of Berg's proposal, Kornberg objected because he also feared that Berg would become the "Pied Piper leading people astray, taking them away from important basic research into this messy field."[39]

Berg, in part frustrated by Kornberg's obvious reluctance, took a one-year sabbatical from Stanford in 1967 for a research post at the Salk Institute in La Jolla, where he could begin to learn about genetic expression in complex tumor viruses free from the expectations of his imposing department chairman.[40]

In the Bay Area's atmosphere of social, political, and disciplinary ferment, Berg was far from the only defector. Many investigators once devoted to fundamental research also considered applied bioscience a reasonable extension. Joshua Lederberg, restless and eccentric, addicted to new challenges, understood earlier than most the potential power of controlling human genes. In his mind, "the research utility of freely moving genes from another species [into the human genome] . . . needs no elaboration." With a remarkable understanding of new trends in bioscience research, Lederberg confidently proposed to the NIH a highly original plan to develop "important practical utilities . . . from the incorporation of [engineered] human genes . . . into therapy for human genetic disease."[41]

In the NIH grant review circles, Lederberg's proposed objectives were taken so seriously that they awarded him, in 1967, the earliest and one of the largest grants to launch a research program dedicated to genetic engineering. One of the brightest stars that Lederberg recruited for his project was a visiting postdoc from Spain, Vittorio Sgaramella, who brought to Stanford a critical but underdeveloped understanding of how to use enzymes to forcibly seal together blunt-end pieces of DNA. His method, however, damaged the structural bases of DNA and rarely worked. In desperate need of advice and guidance, Sgaramella instead found his mentor aloof and often absent, distracted by numerous administrative duties and obsessed with the development of ACME—an early version of the Internet.[42]

Isolated and bored, Sgaramella occupied his time by "attending [biochemistry] group meetings . . . and participating in their discussions."[43]

During one of Sgaramella's presentations, a relatively obscure twenty-two-year-old biochemistry graduate student, Peter Lobban, instantly recognized that the ability to join unlike pieces of DNA would allow bioscientists to build powerful "gene therapies" capable of curing inherited diseases such as diabetes or cancer. A recent graduate in electrical engineering from MIT, Lobban had entered Stanford's biochemistry department in 1967 with a scientist's curiosity, an engineer's appreciation for "inventing new things and applying them," and his generation's propensity for the unconventional. By summer 1969, Lobban had become so frustrated with the distractions of his Ph.D. requirements that he proposed somewhat impulsively to his adviser, Dale Kaiser, that he could improve upon Sgaramella's dilemma—that he could find an enzyme in Kornberg's well-stocked refrigerator to manipulate the ends of two pieces of DNA and make them naturally recombine.[44]

Kaiser, somewhat dismayed by Lobban's inability to finish his original project and intrigued by the novelty of gene therapy, accepted Lobban's alternative proposal on the condition that he dedicate a majority of time to his original dissertation topic.[45]

Berg, upon returning to Stanford in late 1968 from his visiting assignment in La Jolla, was surprised to find that some newcomers were already working on recombining pieces of DNA, a coincidence that would eventually cause some disagreement over origin and competition. But there were also important differences between Paul Berg's goals and the research debuts of graduate students like Peter Lobban: Lobban was an underfunded, underprepared, inexperienced, and overextended student; Berg was a tenured faculty member with a major grant from the NIH that allowed him to bring on numerous postdocs, graduate students, and technicians, all of whom were dedicated to this single research project. The comforts that Berg's team enjoyed also allowed them to be more experimentally ambitious: rather than try to fuse similar pieces of DNA together, such as Lobban proposed to do, they would attempt the unthinkable: they would try to attach unlike pieces of DNA to each other.

As the spectacle of Berg's ambitious proposal unfolded, team members recognized a number of experimental obstacles, three of which would prove particularly challenging. First, they must find a host-vector to which they could attach foreign pieces of DNA; however, the host-vector they were seeking must be small enough to manipulate, yet large enough to accept foreign pieces of DNA. Simply put, single-cell organisms can be too small while mammalian genomes can be too large. Second, they must open the host-vector in such a way that it would accept the foreign pieces of DNA. And third, they must find a way to recombine all of the pieces that was less disruptive than Sgaramella's invasive method. All of these considerations conspired to ensure that a successful recombinant DNA experiment by Berg's team would be a major breakthrough.[46]

Fatefully, Berg had already solved the first problem while on his sabbatical at the Salk Institute. There, he identified a monkey tumor virus, called SV40, which could serve as a host-vector of a manageable size. Identifying SV40 as a vector might have made up the heart of the experiment, but it scarcely meant that success would carry over. David Jackson, a graduate student whom Berg assigned to the second problem, toiled with the problem of cutting vectors so they would accept foreign pieces of DNA, and Bob Symons, a visiting investigator from Australia whom Berg confidently assigned to the third problem, found that Sgaramella's method of forcing recombination of the foreign pieces to the host-vector "caused bad things to happen."[47]

Their early optimism dimmed, coupled with the promise surrounding

Lobban's proposal, Berg's confident laboratory staff became hesitant; they were not wrong, but they knew they were not right, either.

Just up the road from Stanford, UCSF's biochemistry department teemed with intellectual excitement and experimental energy. The new chairman of the department, Bill Rutter, had assembled an incredibly talented group of investigators whose "orientation to higher organisms and humans [made research at UCSF] more serious and practical." Many of the investigators who arrived about the same time as Rutter, such as Harold Varmus, J. Michael Bishop, and Leon Levintow, specialized in the study and purification of enzymes they hoped might be used by those "interested in recombination of DNA." Contemporary observers point to Rutter's decision to select staff with special training in enzymology as the primary reason why UCSF became a central "node in the bioscience research network." While Rutter may have expected that an emphasis on enzymes might improve UCSF's stature within the field, he readily admits that he did not foresee the speed with which UCSF would catapult into the upper echelons of that network, or the source of their success.[48]

Indeed, little did Rutter know that at about the time he had taken control of UCSF's biochemistry department in 1969, his colleague Herbert Boyer had obtained a grant from the NIH to support his search for an enzyme that would allow bioscientists to conduct "genetic surgery." A few months after receiving the award, Boyer stumbled upon a new restriction enzyme called EcoR1, taken from a patient's urinary-tract infection, which could cut or clip DNA with precision.[49]

In July 1971, Tom Broker, another graduate student in Stanford's biochemistry department, overheard his friend Peter Lobban discussing his recombination experiments and suggested that rather than force the two pieces of DNA together using Sgaramella's method, he should try using one of Kornberg's terminal transferase enzymes to make the ends of the two pieces compatible and cohesive. Kornberg, initially hostile to applied research projects that lay "outside [the department's] immediate interests," agreed to share his enzymes when Lobban's advisor and his colleague, Dale Kaiser, defended the project on the grounds that it held "fundamental value."[50]

From the outset, Broker's suggestion proved a huge success. After working through a few early mistakes, Lobban identified an enzyme that less forcefully created "sticky ends" on DNA's broken strands. Ironically, however, Lobban's success fueled new opportunities that would keep him away from his primary responsibilities, many of which had to do with his stature as a graduate student. Kaiser, somewhat dismayed that Lobban had done little work on his dissertation, nevertheless suggested that he share the news of his experiment with his colleague Paul Berg, who was working on a similar project just down the hall.[51]

Peter Lobban never spoke to Paul Berg about his successful "sticky ends" experiment, as his adviser suggested. But he did talk about it with Berg's staff, including his friend David Jackson, who was struggling with the precise problem that Lobban had just solved. Jackson listened intently to the details of Lobban's experiment and sensed "more and more ways in which [Lobban's techniques] were going to be very broadly applicable" to their own experiment and to the broader goal of engineering genetic material. Quite unexpectedly, a chance conversation about "sticky ends" between two graduate students solved the third problem that confronted Berg's group. It remained to be seen how they would—or if they could—solve the last obstacle and open the SV40 vector with precision.[52]

Despite his remarkable scientific successes, however, Lobban faced a professional dilemma. He was running out of time and funding, and he had no assignment for the next academic year; he could conceivably spend the rest of his time at Stanford, but doing so would prevent him from finishing his dissertation or finding a job. His adviser made the decision for him: writing an article would require too much time; the dissertation and the job search would come first. Lobban obediently sent out twenty-two job applications for university positions, fifteen of which requested on-site interviews. He also placed his genetic engineering experiments on hold.[53]

While Lobban continued to work in isolation, Berg's large, young, and ambitious group continued to tap into the leading edge of the bioscience research network. His graduate students gave regular reports on the progress of their work in seminars, memos, and casual conversation, and in turn, received input on what others were doing. Conspicuous among the energetic staff was Janet Mertz, one of Berg's most trusted and active graduate students, who closely monitored developments in the network and kept an especially careful eye on experiments related to genetic engineering. Her training in biology and engineering at MIT, in addition to the unbounded energy that her advisers found a "pain in the butt," made her well suited for such work.[54]

In late 1971, Mertz heard about the EcoR1 restriction enzyme that Boyer had used to make precise cuts in genes. Unbridled by traditional professional protocol, Mertz somewhat naively asked Boyer if she and two fellow graduate students could have some of his EcoR1 restriction enzyme to use on Berg's SV40 host-vector. Boyer generously complied, for reasons that one of Boyer's assistants speculates had to do with the originality of the experiment, its potential utility, and because Mertz was a woman and, thereby, less of a professional threat.[55]

With an air of youthful expectation, Mertz and some of her young colleagues applied the EcoR1 enzyme to the SV40 vector and waited for the reaction. They did not wait long. The enzyme broke right into the vector, opened a specific region, and left the rest of the SV40 genes unharmed. There was another surprise. Quite unexpectedly, when the EcoR1 enzyme broke into the SV40, it left the ends of the vector's DNA strands naturally "sticky" and able to reattach. Amazed at their good fortune, Mertz and Davis immediately told Berg that Herbert Boyer's EcoR1 enzyme solved virtually all of their problems; Berg then called Boyer in San Francisco about "this astonishing thing." Boyer "came flying down [to Stanford] within the hour."[56]

In July 1972, staff in Paul Berg and Herbert Boyer's laboratories put the final touches on three articles that described the EcoR1/SV40 experiment to appear in the same issue of the *Proceedings of the National Academy of Sciences.* Local newspapers gave the experiments a couple of half-comprehending paragraphs the next day. One article said that bioscientists at Stanford and UCSF had "broken the species barrier" and wondered naively about curing cancer.[57]

The public may not have understood the first recombinant DNA experiment, but colleagues in the biosciences most certainly did. The effort spawned a prolific brood of visitors to Berg's lab, one of whom was Stanley Cohen, a colleague of Berg's in Stanford Medical School's clinical pharmacology research program, who stopped by to see what the commotion was about. What he did not see would prove formative in the future development of more powerful genetic engineering techniques.

"You won't be able to clone it," said Cohen, who instantly recognized that Berg's decision to use the SV40 monkey tumor virus as a vector prevented replication—and practical application.[58]

Cohen's observation posed a huge conundrum for Paul Berg. On the one hand, no practical health benefit could come from his experiment; he and his staff had merely shown that it was possible to attach unlike genes together. The limits of their experiment, as Cohen noted, lay with Berg's early decision to use the SV40 virus as a vector: viruses live off host cells; therefore, they cannot carry an engineered gene into humans. In the context of what Berg called "the new social conscience of the era," therefore, Cohen's remark contained an intriguing experimental possibility of curing disease through genetic engineering: "We were all of a like liberal mind," noted Berg, "and . . . felt that ethics and responsibility in science was now important. . . . There was a sense of wanting to do the right thing."[59]

Many investigators, in addition to Berg, looked upon genetic engineering as the bioscience idiom for "the right thing" because it was pure research, it had practical value, and it was idealistic. An assistant in Boyer's lab at UCSF recalls the experiments: "By 1968 or 1969, I was interested in thinking about some way to start manipulating genes. . . . There was somehow a feeling in the field . . . to start thinking about tying different kinds of DNA molecules together. It was something I had on my mind for a long time. And others did."[60]

Indeed, as a Stanford graduate student in the biosciences understood it, "usually scientists do experiments that are useful for other scientists. . . . But this would be one of the few times a scientist really had an opportunity to do something for the general public." Another, awed by the promise of genetic engineering, exalted in Berg's successful experiment because "it meant that you could do anything with the genes of any organism. . . . I think that the potential for affecting . . . agriculture and medicine was obvious to us—I know it was."[61]

It was clear to everyone, including Berg, that genetic engineering could "affect" agriculture and medicine, maybe even feed the masses or cure the sick. But how? Berg wondered. He agonized for months trying to figure out how to engineer the human genome; then a telephone call from an old friend and colleague, Bob Pollock at Cold Spring Harbor in New York, convinced him to reconsider the experiments—especially cloning—until more stringent controls had been established. "Why are you doing this crazy experiment?" asked Pollock rhetorically. "Could [you] create a new cancer that was infectious for humans?"[62]

Berg hesitated because, in addition to the fearful prospect of creating an uncontrollable biohazard, he also took note of the irresponsible climate in which genetic engineering research was being conducted. He was not the only one to notice, or to have such qualms. Just months after making important contributions to one of the most revolutionary bioscience experiments of the era, both Lobban and Mertz left academic research, in part to get away from the competition, intolerance, and runaway ambition that drove the newest generation of bioscientists. A graduate student in Herbert Boyer's lab recalls: "there was quite a bit of intra-departmental rivalry at the time. . . . I remember we felt the competition. . . . They put out these little newsletters, like 'The Midnight Hustler,' talking like we were sports teams or something." Another offered a more caustic critique: "God forbid you were . . . competing on the same project with a postdoc in [another] lab. . . . [It was] hardly an environment to foster interactive, collegial collaboration."[63]

Berg reflected on the potential for this new generation of bioscientists creating runaway biohazards at a feverish pace. It seemed that investigators in this new era could justify any applied research experiment, even if it required unsafe and unreasonable experimental risks. He saw graduate students suck viruses into pipettes using their mouth. Others dumped enzymes down the drain with impunity. He even witnessed far too many working with "huge amounts of radioactive phosphorus" without bothering to put on gloves. And he heard about a mistake that a British laboratory had recently made with smallpox that had led to some deaths. To Berg, these unacceptable experimental practices bore the unmistakable imprint of youthful idealism run wild. With great reluctance, he made the difficult decision to "put recombinant DNA experiments on the shelf" and reconsider them in the context of hypothetical biohazards, primarily because too many of his colleagues seemed "selfish . . . and would pursue [cloning] hell-bent; no matter what anybody said, [they] were going to do the experiment."[64]

To enough investigators in this new era, however, "hell-bent" seemed like a reasonable way to pursue genetic engineering, especially if their work might cure disease.

Embedded deep within the disagreement between the two generations of researchers was the perception of retreating basic bioscience experimentation and the rise of relevant bioscience research. For the two decades following World War II, bioscientists at Stanford, UCSF, and Berkeley pursued religiously, and with few interruptions, fundamental knowledge. It was widely assumed by this group that the field would advance most rapidly if it avoided external influences—no other experimental approach was considered workable or efficient. In contrast, they also believed that any relationship with relevant concerns would divert attention away from emerging "truth" and would render unreliable all experimental outcomes of this nature. Their highest ideal—the supremacy of basic research—informed individual investigators, medical school programs, and even the entrepreneurial pursuit of federal patronage.

Their search for fundamental bioscience truths in the immediate sense was irrelevant—that much was undeniable, even for those conducting the experiments. But bioscientists in this era believed that with enough time, money, and resources, they could learn almost everything. What they wanted was the public's unconditional patience and financial support, and for nearly twenty years they got it; but that patience ran out by 1967. Then, as the above examples show, certain groups of bioscientists began to accommodate the public's newest concerns, which set the stage for an unusually intimate and productive application of existing fundamental knowledge for practical purposes.

No sooner had the world learned of the epochal moment for applied bioscience research—Berg's successful experiment combining pieces of DNA—than a fatal elision took root between those who supported genetic engineering research and those who opposed it. The differences that emerged between these two sides during this period would bedevil the field and, more broadly, suggest some of the difficulties endemic to applied bioscience research and the idealism and competition that it spawned. Somehow, both sides perceived their antagonists as obstructing, corrupting, or indulging in the experimental order.

In reality, the disagreement was not about experimental differences or competition, but about the inconsistencies that resulted when the boundaries that once separated pure and applied research became blurred, which rendered the entire conflict meaningless. The battle between pure and applied bioscience was all but over. A new boundary had yet to be determined. The battle for advantage in the age of genetic engineering was about to begin.

Chapter 8
Cetus: History's First Biotechnology Company

Sometimes I couldn't tell if we were the rearguard or the vanguard.
—Ron Cape, founder and president of Cetus

The first biotechnology company started with a machine. Not just any machine; it was a bioengineering machine. It could induce mutations in a massive vat of organisms. It could identify strains for potency, reproducibility, morphology. It could make clones. No one at the beginning knew what to do with the machine, or even how it might be used, but they could imagine without difficulty the heights they could achieve—not only scientific distinction and not merely survival among commercial giants.

The long-term prospects were indeed fantastic, but the short term was plagued with uncertainty. As the idea began to take shape and a company began to emerge, the founders struggled between new versions of old tensions: should the company follow the more traditional route, producing and then selling the machine as an experimental tool for university research, or could the company use the machine to actually make a bioscience product, something that had practical and commercial value?

In its most familiar form, the central challenge to starting the first biotechnology company had to do with the very nature of what the biological sciences had been, versus what the biological sciences could become. But the source of the disagreement went much deeper, involving conflicting attitudes about whose insights should hold more weight: science, capital, or public good? Whether the new generation of bioscientists were quixotic, or blindly pursuing an alchemist's dream, they stood alone like Janus, both in the rearguard and the vanguard. This battle—between bioscientists, capitalists, and the public—would determine the fate of the first biotechnology company, and the shape of an industry forever.

The Bioscientist and the Machine

In 1960, Donald Glaser won the Nobel Prize for his invention of the bubble chamber. It was an experimental tool that high-energy physicists used to literally see everything: the interior of atoms, the structure of particles, the composition of matter; it was so far-reaching that it ran through the settling of modern nuclear physics for years to come. In his effort to make a scientific field more efficient, Glaser had shown himself capable of finding pragmatic solutions that could carry him and an entire scientific field into uncharted frontiers, and beyond. By all accounts, it was a remarkable performance. At only thirty-four years old, he had reached elite status within physics, and in the public's imagination.

But Glaser found the limelight uncomfortable, and worse, his scientific field no longer relevant. So he quit physics, on the grounds that pure research—conducted by swarms of students and faculty using massive and incredibly expensive machines—had little practical value to society, or worse, contributed to destructive warfare. Then he and his young wife moved to Boston, and he spent a semester sampling introductory courses in the biological sciences at MIT and then another semester as a postdoctoral fellow at the University of Copenhagen to study microbiology. By the time that Glaser returned to Berkeley a few years later he was ready to restart his scientific career, but this time, in a scientific field that gave something back to life.

Where exactly was Glaser going, many of his colleagues wondered in dismay? Debates about Glaser and his decision to leave a scientific field that he had conquered only to start over in another tore at his friends and UC administrators, and tore departments apart. But Glaser had more in mind than simply avoiding heightened professional expectations; he was determined to make a scientific field more efficient, more practical, and more relevant.[1]

While studying microbiology in Denmark, Glaser witnessed firsthand "a monotonous experimental method incapable of seeing beyond pure research." He saw rows of microbiologists sitting at their lab benches in isolation, quietly spreading cells on nutrient agar in petri dishes. He watched them wait, sometimes as much as a day or two, for individual cultures to incubate and grow into cell colonies a few millimeters in diameter. They would peer through their microscopes and search for colony "fingerprints"—a recognizable shape or size, signs of mobility on the agar surface, or sensitivity to various applied stimuli. Even more stunning, from Glaser's perspective, was that the most advanced experimentalists approached their research in the simplest manner, even if complex and ramified in detail. They would pluck a suspicious organism from a vat of colonies, re-culture it in a variety of liquid suspensions, and then conduct detailed biochemical, serological, and microscopic staining techniques to produce an effect only partially understood at the time. Most of the experiments that Glaser saw, much to his wonder, took anywhere from two days to two weeks to complete, and in the best case, the highest trained technicians could identify just three out of four cultures with any reasonable degree of certainty. Here was an opportunity to make an entire scientific field more effective, more useful, and more practical; to introduce, in short, the principles of engineering to the biological sciences.[2]

Glaser's quest to remake the biological sciences started modestly enough, when he designed, simple in outline and useful in its application, a machine that he called "the dumbwaiter," which stacked eight one-meter-square trays, each holding ten petri dishes at a time. To make the dumbwaiter operational, however, he integrated it with a much more sophisticated machine that he built called Cyclops, which illuminated the petri dishes with a light from below so that an overhanging camera could take time-lapse photographs of the growing cultures. To permit experimental manipulation, Glaser added to each tray a series of pipettes that administered to the growing cultures a range of chemicals, such as amino acids, penicillin, or vitamin B, in order to induce mutations or other external variations. Then, to allow for data collection, he attached a computer to record the behavior of each individual colony. Less dramatic, but of considerable importance for advanced microbiological research, Glaser included an intricate mechanical hand that used tiny quartz-rod fingers to pick up specific colonies and lay them down on other trays for further concentrated study.[3]

Without question, Glaser's "dumbwaiter" and its many variations constitute an engineering marvel, but no design better exemplified Glaser's leaping mind and artful talents than his machine that made clones: the Lazy Susan. Anticipating the technology later used in inkjet printers, Glaser built a machine that generated drops that contained, on average, a single bacterium from which an experimentalist could then grow clones. Glaser found it surprisingly easy to get his initial design to produce drops that contained bacteria; the central challenge that he faced, however, was to create a machine that produced drops that each held *one and only one* bacterium. To combat this problem, he rigged a laser beam to shine light on each suspended droplet as it formed, and then attached a computer scanner to analyze the light that passed through each particular droplet. If the computer recognized that the laser beam reflected only one bacterium in the suspended droplet, then the machine would literally drop the individual bacterium into petri dish, ultimately producing a sheet of colonies, each one guaranteed to produce a clone. However, if the computer recognized that the laser beam that had passed through the drop refracted, or if the computer recognized that the beam reflected two bacteria in the drop, then an automatic electrical charge would push the unwanted drop away. From there it was relatively simple to apply his other inventions, such as Cyclops, which photographed cloned colonies for advanced morphology studies.[4]

Glaser's machines were apparently compatible, in his mind, with his belief that he could make the biological sciences more useful to humanity. Inconsistencies notwithstanding, Glaser's technological remedies for the biological sciences reflect the unmistakable signs of a distinctive engineering genius and a mastery of design seldom duplicated. Simply put, Glaser's series of machines that he called "a screening system"—the dumbwaiter, Cyclops, and Lazy Susan, as well as "baby counter," Roundabout, Candid Camera, "colony picker," and a host of other machines—embodied the principle of bioengineering that lay at the heart of the coming scientific and industrial revolution. The only method that might have been more efficient than using Glaser's bioengineering machines to *find* a desired organism would be to develop a bioengineering technique to *make* a desired organism. But in the mid-1960s that was a theoretical way of doing bioengineering, and it would happen first in academic laboratories far off in the future, or so thought a handful of bioscientists, such as Paul Berg and Stanley Cohen at Stanford, or Herbert Boyer at UCSF.

Many more scientists knew about Glaser's bioengineering screening system than knew about developing genetic recombination methods, and they found the former inspiring. An official at the Centers for Disease Control speculated that a bioengineering screening system would use just "one-third of [a specialist's] time and cost half as much." That was not nearly ambitious enough, said another: "it will reduce a typical eight-man-hour task to about two hours and save as much as $18,000/man/year." Casting aside all restraint, a science writer boldly declared that Glaser's bioengineering system would make advanced academic training in the biological sciences obsolete. Of those who saw the coming of the new biosciences, perhaps the most sober assessment came from Robert Angelotti of the FDA, who conceded that a "ready-made market" already existed for bioengineering. When asked to elaborate, Angelotti tried to temper his obvious enthusiasm by repeating a theme that had become popular in the late 1960s: "the need exists."[5]

Need perhaps, but academia approached Glaser's new bioengineering screening system with paralyzing indifference. Most biological scientists could not imagine an experimental approach that could render more reliable discovery. Arthur Kornberg, biochemist and Nobel laureate at Stanford, spoke for many when he warned that bioengineering would one day "lead everyone astray." It would become fashionable in later years to dismiss opinions such as Kornberg's as tragicomic evidence of academia's quaint, ideologically hidebound fear that bioengineering would spell doom for less popular bioscience subdisciplines, or taint the cherished objectivity upon which the profession rested. But such concerns should not be so summarily dismissed. Glaser's machines had passed only preliminary tests, no one had yet considered the prospects of biohazards, and most academic laboratories did not have the space for such large and expensive machines or the money to purchase them. Moreover, although it was only obscurely visible at the time, there was more than a wisp of truth in what Kornberg feared: bioengineering would one day circumscribe many academic bioscience programs and become virtually indistinguishable from bioengineering practiced in commercial industry.[6]

The Venture Capitalist and a Dream

The isolation of Glaser's microbial screening system from academia still left plenty of room for a young venture capitalist named Moshe Alafi to think one step beyond: the application and consequences of applied research. Even as a child, Alafi always saw and did things a little bit differently. Born sometime in 1940, Alafi grew up in Baghdad, the son of a Jewish upper-class merchant. He attended the distinguished French college-preparatory school Alliance Israeli Universelle, where he enjoyed a comfortable and privileged education that was walled off from an otherwise violent world. He was torn, unable to make sense of the conflicting circumstances that defined his community: conspicuous wealth amid religious conflict. Confused, he favored instead risk-taking decisiveness—taking a side even before all the evidence is in—as a way to move ahead of difficult issues. He made decisions, and then, just as decisively, would change his mind. His fluid manner and uncompromising approach touched everything he did, including a unique sense of bioengineering and its enormous commercial capabilities.[7]

After graduating in 1957, Alafi immigrated to the United States and enrolled at the University of California, Berkeley, as an undergraduate in physiology. He enjoyed the spirited freedom of Berkeley life so much that he entered graduate school, switching fields of study within the biological sciences, from physiology to biophysics and then back to physiology. Characteristically, Alafi treated his college education more as a hobby than an intellectual pursuit. Indeed, while a graduate student he started a hosiery-store chain, which most certainly stood out in Berkeley, much like the double-breasted, pinstriped blue suits that he occasionally wore to class.

Just as suddenly, Alafi quit graduate school and started a company, Physics International, which sold nuclear medical equipment. One year later, he shifted the company's primary market to testing nuclear Minuteman missiles, for which the U.S. government paid handsomely. Those who knew him during this time, whether in business or personally, admitted that his pace and tack was a large part of his appeal. University of California Regent Ed Heller respected Alafi enough to invest a half-million dollars in his early business ventures. The bohemian poet Lawrence Ferlinghetti enjoyed Alafi socially, and often gave him keys to stay in his Big Sur cabin. And Ed Carter Hale, chairman of Neiman Marcus, always invited Alafi to his intimate cocktail parties. Alafi thoroughly enjoyed his life as a local celebrity, and more so when Physics International went public in 1964. But he had no desire to work for anyone else but himself, particularly for a publicly owned company. So Alafi sold his share of Physics International and began to drift, searching for another big venture.

For anyone other than Alafi, it would have been a poor time to change careers. The quickening of America's war against communists in Vietnam and against poverty at home had created, by the mid-1960s, a surreal and inherently unstable economy. The steady growth of the nation's financial markets—which had been going on since Kennedy's first year in office—had taken an abrupt turn: inflation and unemployment levels were rising, real wages and consumer spending was falling, the nation's GNP was stagnant, and business startups, long recognized as a powerful countercyclical tool, had slowed most dramatically.

Yet, so far as the economy was concerned, things could hardly have been better in California's Bay Area. The roots of the boom that delivered opportunity to entrepreneurs such as Alafi can actually be traced to 1955, when Shockley Semiconductor planted the seed from which grew a plethora of electronics spin-offs, including but certainly not limited to Fairchild, Intel, and Tandem Computers. Meanwhile, Varian went public in 1956, followed by Hewlett-Packard in 1957. Then, in the wake of *Sputnik,* a Niagara of federal funding for high-tech weaponry and gadgets flooded local firms with lucrative contracts, of which Alafi's Physics International was just one of many that profited. Before long, a fabulous amount of wealth had flowed into the hands of a relatively small number of local entrepreneurs. Meanwhile, Congress, searching for an opportunity to shift some of the burden of economic growth from the public sector to private markets, passed the Small Business Investment Act in 1958 to entice private investment in small start-up ventures with tax breaks and matching funds up to $300,000. In reality, the SBIC had very little direct impact on business development: the program's $5.2-million budget was smaller than that of the Office of Coal Research, its staff of thirty-one was less than one-tenth of the Federal Crop Insurance Program, and all tax breaks and credits passed through the hands of the investors before the capital reached a startup's ledger. For all the modesty of this enabling legislation, however, the SBIC showed its might a few years later when it brought the new class of Bay Area industrialists into a formal and professional investment activity. By the mid-1960s, the core of venture capital was born.[8]

In 1965, Ed Carter approached Alafi and asked if he would like to join as a general partner in his venture capital firm, Murray Hill Scientific Investment Company. Naturally, given his proclivity for risk and adventure, Alafi accepted the offer. Also quite naturally, while most venture capitalists at the time were leaping into computer technologies, Alafi deliberately looked for other business opportunities.[9]

Alafi came to know Donald Glaser at about the same time he became a partner at Murray Hill. They met socially first, probably at a neighborhood cocktail party, and then became friends as their two families spent time around Alafi's swimming pool. For Glaser, the introverted tinkerer, Alafi's cosmopolitan ways seemed the perfect counterpart, and a bond based on mutual respect soon formed between the two men. It was at one of these informal gatherings sometime in summer 1965 that Alafi learned about Glaser's machines.[10]

Alafi did not lack for evidence that Glaser's bioengineering machines had enormous potential, so he skipped the customary due diligence that marks venture capital and pressured Glaser to start a company as a partnership right away. He asked Glaser to see with his seasoned scientific eye the lives that they could save, the mouths they could feed, and the illnesses they could cure. As a further emolument, Alafi carefully pointed out that the medical-diagnostics market alone easily surpassed $50 billion. Then Alafi asked Glaser to imagine controlling even larger markets. It would take Alafi more than a year to convince Glaser to start a company, and then, said Glaser, only if Bill Wattenberg, a faculty member in Berkeley's computer science department who had designed a prototype for a personal computer, could join the partnership.[11]

With their consent in hand, Alafi plunged ahead with a remarkably simple business strategy: turn the academic research of his two partners into commercial products. Since Glaser and Wattenburg never swayed from their work, it seemed as if the venture would remain productive forever, and a biotechnology company might become a reality at stunning speed. They quietly used a portion of Glaser's NIH and NSF grants as start-up capital, and then incorporated the business in August 1966 as Berkeley Scientific Laboratories (BSL), renting a small office at 2229 Fourth Street in Berkeley. Then they sat back and waited for marketable products to pour out, but none did. No one, it seemed, had any idea if a bioengineering company could earn revenues, or how.[12]

Haste had precluded serious thought, but just as harrowing, Alafi had woefully underestimated the hostility that many in academia still harbored toward applied research—especially research for commercial possibilities. Out of respect, none of the faculty at Berkeley openly criticized Glaser—their Nobel laureate—but they did not hesitate to turn the full force of their fury on his partner, Bill Wattenberg. His department assigned him a grueling teaching load and endless committee work, and then the administration dismissed his tenure application on the grounds that he had produced an insufficient amount of research. By summer 1968, the most committed of the two scientists running BSL had quit. That fall, a dejected Glaser went to Alafi and told him that he would give up on BSL too. He felt overextended, he explained, having lost a partner and colleague, the respect of his department, and a substantial amount of his personal savings; his grants from the NIH and NSF had run out and his request for renewal had been rejected, and to make matters worse, his wife no longer appreciated his work habits and was going to leave him. He still believed in his bioengineering system, of that Glaser assured Alafi, but he would not and could not run BSL on his own.[13]

Breaking Ground

It was Alafi's good fortune to have at about this time two visitors. Ronald Cape and Peter Farley were remarkably similar in background and outlook, and would become linked not only because both came to the Bay Area in search of fame and fortune, or because they incorporated as Cape/Farley only weeks after they met each other in Alafi's office. To begin, they were young—in their mid-thirties, with Cape just five years older than Farley—and enormously sociable, making friends with remarkable speed. Both spent years in graduate professional programs: Cape earned an MBA from Harvard, which he found "superficial," then took a Ph.D. in biochemistry from McGill, followed by a postdoctoral fellowship in molecular biology at Berkeley; Farley received his M.D. from St. Louis University, but realized that "he really got his kicks" studying finance in the MBA program at Stanford. Along the way, they both entered occupations they found unsatisfying: Cape wanted little to do with his family's cosmetic distributorship in Montreal and no part of his present occupation as an "irrelevant" academic; Farley left his private medical practice in Honolulu because he could no longer tolerate "the number of people going into hospitals simply because they had nobody to take care of them at home." And both had advanced knowledge of the life sciences—in particular, knowledge about medical science—that was fairly prodigious, and they had a knack for conveying it in a style that nonscientists found accessible and intellectually thrilling. There was, however, one overriding difference between them. By force of training, Cape was inclined to defer to science and the insights of its practitioners, much like Glaser; by force of habit, Farley deferred to business and deeply trusted financial results, much like Alafi.[14]

Cape and Farley's experiences, in graduate school and in the workplace, mirrored a generational shift. Alafi sensed this: to him, Cape and Farley seemed to have all of the right qualities to run a bioengineering company, so he urged Glaser to try again. Glaser, however, still held all the high cards. He would relent, he said, but only if he could stay as far away from the business as possible. Alafi gladly accepted his condition and then introduced Cape and Farley as the two people he thought should run the company. Glaser found them both competent, as Alafi said, and then he played his hand too strong. Glaser asked that this "leadership team [Cape and Farley] . . . consider producing a product . . . with a longer development time rather than rushing to the market with a compromise system."[15]

"Time," Alafi shot back, "is not our friend"—an indication to his neophyte partners that they must move quickly to capitalize on their first-mover advantage in bioengineering. In reality, a surprise competitor did not exist and had little basis in fact. Richard Sweet, a professor at Stanford University, had invented a machine somewhat similar to one of Glaser's scanners, but he had no interest in business. That left Collaborative Research, in Waltham, Massachusetts, and Green Cross, a Japanese company modeled after the humanitarian organization Red Cross, as the only two companies manufacturing screening machines at the time, and nothing coming out of either of the two companies remotely suggested that they posed a competitive threat. Alafi's warning nevertheless fell on receptive ears. The businessman's acumen would trump that of the scientist's.[16]

As the four partners and one employee got down to work, a sense of high excitement took over. Morale was high, the hours long, the dedication total. On October 1, 1971, Alafi scrambled together a basic partnership agreement followed by a skeletal framework for a company, one in which he, the visionary that he thought he was, would serve as chairman of the board; Glaser, naturally, would lead the scientific advisory board (SAB). Ronald Cape and Peter Farley, Chief Executive Officer and President, respectively, would thereafter "constitute the Executive Committee" in charge of "promotional responsibilities critical to the company's success." Then, in late December, Cape and Farley signed leases and began moving used office equipment into 851 Dwight Way, just off the west side of the Berkeley campus, and Glaser's bioengineering machines into a recently re-fabricated site at 600 Bancroft Way, near the Berkeley marina.[17]

Cape and Farley had already begun thinking about a name for the company. Since they had not yet decided on a market or a product, they wanted a name that sounded well born and well placed. They grappled with abstract combinations of syllables taken from biology and technology, until late one evening the single employee, Cal Ward, told them a fishy story about a shark attack just off the Pacific coast. It was a "whale of a tale," said one dismissively, but the comment startlingly captured their imagination. A whale indeed, but they wanted something bigger, infinite even, and universal too. Cape looked up into the sky and pointed out Cetus, the cluster of stars in the shape of a whale. Here at last, in the guise of a company name, was a symbol that conveyed grand undertones for what they hoped their venture would become.

They called the company Cetus Scientific Laboratories.[18]

In February 1972, just four months after forming partnership and one month before formal incorporation, Cape unveiled the Cetus business plan. He had little time to draft a carefully crafted document, or iron out all of the wrinkles, but its heart conveyed a simple message: "To introduce sophisticated systems and instruments to the practice of medical and biological research." In a masterful way, Cape capably avoided a comprehensive outline of the company by turning ambiguity into the company's strength: "the instruments and systems currently under development embody many secret and proprietary contributions. In order to maintain security, [the Plan] will not disclose in detail Cetus instruments and systems." After advancing a few ideas about their market space, which he dismissed as practically self-evident, Cape carefully noted that the principal risk for the company came from an unlikely source, something he called "old friends—the customer's inertia and existing habit patterns." Naively, however, Cape argued that the company would eventually overcome the habits of "old friends" because markets always respond to superior products like Glaser's microbial screening system. As to the finer point of how, exactly, would Cetus earn revenues, Cape took no chances and proposed two entirely different scenarios. The first he described in a straightforward manner, designated as "System A," in which Cetus would become a vertically integrated company that manufactured and sold Glaser's machines as experimental and medical devices. Tellingly, there appeared buried deep in the text a second and completely separate business strategy, referred to as the "Mutant Search Program." It came, ironically, from the partner with the least amount of experience in experimental biology—Farley—who nevertheless showed a deeper understanding of bioengineering's destiny than his more scientifically sophisticated partners: "an opportunity exists for the application of Cetus technology in conjunction with today's understanding of molecular genetics." The revolutionary nature of the Mutant Search Program—specifically, how to bioengineer new organisms with Glaser's machines in a way that earned revenues—lent point to their reluctance to commit to this model.[19]

One month later, on March 27, 1972, the four founders of Cetus crowded into the majestic corner office of the law firm Heller, Ehrman, White and McAuliffe on Montgomery Street in downtown San Francisco to meet with a group of potential investors. Alafi opened with a stirring confession: the gathering had been selected for their comfort with creating a revolutionary industry. Then he introduced Donald Glaser, who quickly sketched his bioengineering system on an easel and then withdrew, but not before his presence impressed upon the gathering that Cetus had a Nobel laureate as a partner. In a coy way, they proposed the medical and biological research markets that Cetus could revolutionize: antibiotic diagnostics was a $75-million market; clinical diagnostics a $350 million market; antibiotic sales, $660 million; antibiotic contract research, $425 million; and so on. Doubtful as a matter of starting a business, any of the markets seemed eminently sensible. It had by the end of the meeting begun to dawn on the investors who had gathered that spring morning that no matter which direction they chose for bioengineering's first venture, a 5 percent share of any one of these markets might be worth millions. The Cetus founders were overwhelmed by unequivocal interest that the investors had for their amorphous business model, which fatefully made the specter of raising capital to start a bioengineering company all too easy (see Table 8.1).[20]

An Elusive Business Plan

No sooner had the ink of the investors' signatures on the financing deal dried—and well before the Certificate of Incorporation had been officially amended to include the sale of stock—than the founders began to "scale up operations." Cape and Farley took the first step when they opened the 1971 *International Microbiology Society* annual industrial report that listed companies that used microbiology by largest revenues earned: Johnson and Johnson, Squibs, Baxter, Abbott, and as far ranging as ConAgra Foods to Miller Beverages, and as notable as IBM, Johnson & Johnson, and Bayer. Not surprisingly, Cape and Farley contacted these companies in a confident and casual manner too. Farley never shied away from cold-calling a top executive, on occasion Cape and Farley

TABLE 8.1. CETUS FIRST-ROUND FINANCING

Purpose of issue	
Fund System A	$550,000
Exploratory capital for Mutant Search Program	250,000
Operating costs (two years)	400,000
Verification testing	150,000
Expenses (banking and legal fees)	75,000
Cash reserve	575,000
Total	$2,000,000
Capitalization	
Series A Convertible preferred (200,000 authorized shares)	$2,000,000
Common stock (2,500,000 authorized shares)	60,000
Total	$2,060,000

road-tripped in their hippie VW van to literally knock on doors, but most often, they wrote letters of introduction, making sure their Nobel laureate Donald Glaser inked legitimacy to their proposal with his signature. With heroic assurance, urgent rhetoric, and appeals to idealism and capitalism, Cape and Farley made the choice plain: Cetus would either manufacture and sell Glaser's bioengineering system (System A), or Cetus could be hired as a service in which they would bioengineer preordered microorganisms in-house using Glaser's system (Mutant Search Program). David Taft, vice president of research at General Mills, still recalls years later his introduction to the Cetus promise: "What they were doing, what they were talking about, was really exciting. Everyone wanted to know more. I know I did."[21]

"Old friends" may have been attracted by the bait, as Cape and Farley hoped, but their reluctance was not truly anticipated. Eighteen different companies welcomed Cape and Farley for their presentation of Cetus. Virtually all of the executives they contacted confessed "astonishment by what Cetus had to offer." All expressed sincere interest in both System A and the Mutant Search Program. And when pressed to sign a contract, all eighteen companies said, no. From their perspective, Cetus was a new and relatively small company, front-loaded with a Nobel Prize winner and an extremely talented leadership team of MBAs and M.D.s, boasting of a secret technology. And they feared that this new bioengineering company might put entire industries out of business, an impression the founders did not try to dispel. Contrary to the received opinion in later years, "old friends" such as the pharmaceutical industry did not necessarily move too slowly when they confronted for the first time the prospect of bioengineering; instead, they expected too much.[22]

Of those companies that Cape and Farley contacted, the most interested was Schering-Plough, a mid-level pharmaceutical company. The extensive negotiations between the two companies may have been overly premature, considering Cetus had not committed to System A or the Mutant Search Program, but Schering was nevertheless paralyzed by a self-inflicted scientific wound. Virtually all of the company's revenues came from micromonospora—a rare bacterium found only in Lake Heviz, Hungary—that secreted a powerful antibiotic called gentymycin, often referred to as "the antibiotic of last resort" in medicine. Schering scientists had dedicated virtually all of their time and resources extracting gentymycin from micromonospora that at best nibbled at the edge of its potential—a lengthy two- to three-year development process that produced an antibiotic that had, in recent years, lost much of its potency. As frustrated as the Schering scientists were the Schering accountants, who considered the $100-million gross revenues gentymycin generated each year an underperformer, or better, "the antibiotic of last choice." Even worse, the company's patent on micromonospora was running out. When Cape and Farley arrived to tout their bioengineering as a scientific solution, Schering's executives had already concluded they had a pressing scientific problem.[23]

It was in this foul corporate context that Schering's Director of Microbiological Research Dr. Marvin Weinstein invited Don Glaser to tour the company's experimental laboratories. As hoped, Glaser identified the company's problem, and much more besides. In a windowless, dingy laboratory in the heart of industrial Trenton, Glaser saw swarms of microbiologists engaged in an "amusing 'hunt and seek' approach" to research and development that he thought "lagged academic microbiology." He found himself face to face with "comatose technicians . . . using toothpicks . . . to pick at colonies, searching for . . . something that looks like [it] *might* belong to the genus micromonospora." "The remarkable fact," continued Glaser, "seems to be that when one [of the technicians] finds an antibiotic produced by an organism there are actually 5 or 6 other antibiotics present, up to 15, coming from the same culture, and the highest producing strains are also the ones that are more unstable so that commercial batches often have to be restarted." In an urgent message to his colleagues back home, a normally reticent Glaser summed up his tours of Schering as "ripe with opportunity."[24]

Pete Farley quickly followed Glaser to New Jersey and dominated all matters great and small. "Cetus' potential gift to the world," Farley proudly declared, "will, for all practical purposes, revolutionize the antibiotic end of the drug industry." The math was simple, intoned Farley. "Schering technicians carry out 400–600 drug assays per day," while a "single tray used [in the Cetus bioengineering system] holds 100 assays each," which, according to rough calculations, would increase the total number of bugs tested each day by a factor of 10^5. Farley was just getting started. Cetus would, if Schering executives so wished, use its bioengineering system to discover new antibiotics among the bugs that untrained technicians discarded. Then, relishing the power that he thought Cetus would soon possess, Farley issued a bold ultimatum: "we will choose the course that seems to us to produce the most dollars down the road for Cetus. Very simple, very easy. . . . We will hold up the entire drug industry, essentially put the technology up to the highest bidder."[25]

Privately, Schering officials did not rejoice in Farley's presentation. As the executives at Schering perceived it, neither failure nor fulfillment inspired total confidence. Conspicuous among Farley's presentation was the blatant disregard for Schering's pressing need to improve the toxicity of its gentymycin strain. But more worrisome, did Cetus really possess such a powerful bioengineering machine? If not, was this a sinister attempt to steal their patent secrets on micromonospora? On the other hand, was Cetus negotiating with Schering's competitors, as Farley implied? Even more consequentially, if bioengineering worked as Farley said it did, what would happen to Schering and the rest of the pharmaceutical industry? Taken aback by Farley's over-the-top performance, Schering executives stewed about their dilemma for months. Then, in spring 1972, they reopened negotiations with a direct offer: they would consider buying the bioengineering machines, or they would hire Cetus to bioengineer improved strains of gentymycin, but only if Cetus shared details about how the entire system worked.[26]

Schering's insistence on a precise description of bioengineering had all the appearances of a reasonable request. It also contained sinister implications for a specialty producer dependent upon intellectual property. From Cetus' perspective, to give away secrets would give away the company, but cash flow problems favored telling Schering everything. For a company less self-assured, the constant rejections and subsequent offer by Schering might have been enough reason to sacrifice the future for immediate gains, but Cetus executives did not lack self-confidence. So they took back their offer and cast about for more amenable audiences. The decision to protect company secrets was reasonable and would one day become a common practice for the industry, but it did nothing to solve their pressing cash-flow problems.

At this moment, Cetus stood on the shore of a financial rubicon. One year earlier they had plunged in deeply to start the company, but now they longed for more shallow waters in which to establish a bioengineering company, or at least decide on what a bioengineering company should do or be. Throughout all of their earlier discussions, late-night brainstorming sessions, and even in their business plan and presentations to venture capitalists, they adamantly refused to choose between the two available markets. It was not a decision they wanted to make. They even made up their own business model that justified their timidity: "don't put all of Cetus' eggs in one basket, in our own heads, or in anyone else's." As Cape recalls, "everybody was looking for a model but it became quite clear that there was nothing for us to follow." Their dream of starting a company that would lead an industrial revolution, they knew, was damned if they chose a direction and damned if they didn't.[27]

Fired by desperation, Alafi returned to Schering, the company that expressed the most sincere interest, and offered a wholly new and creative proposal. If Schering truly expected bioengineering to fail, then to prove his sincerity Cetus would use their machines on micromonospora, charging a royalty according to how much gentymycin they found through bioengineering.[28]

Certainly, it was unusual for a venture capitalist such as Alafi to bypass short-term revenues, but his strategy of negotiation derived not from the promise of good faith he had made, but from the arithmetic of his expectations. Needing a contract—*any* contract—as a starting point, he anticipated using Schering as the example that would force the hand of other pharmaceutical companies to act. He and the other founders also intended to fulfill any obligation it had to Schering, but they said nothing about how they would approach other pharmaceutical companies. Privately, Alafi knew that the present cash-flow problem was not as urgent as it seemed because Cetus had at their call a host of investors ready to participate in a second round of financing. From Schering-Plough's perspective, the company had been profitable for almost a century, so executives there knew a good deal when they saw one. They needed proof that bioengineering worked, said Schering's attorneys, but they did not wish to take on a truly complex project, or provide support for a company that might one day put them out of business. Rather, they would "accept in principle, the concept of a fee for using [Cetus as] a service." Barely masking their enthusiasm, Cetus immediately signed a contract on 9 July 1973, with only a few minor revisions.[29]

With that, by way of a desperate offer, Schering forced Cetus to become a bioengineering company.

In retrospect, the contract hammered out between Schering and Cetus must surely rank among the vaguest in industrial law. Its principal agreements lacked detail. For instance, it stipulated that Schering would send Cetus strains of micromonospora, but the contract did not specify the quality of the strain they sent. In absentia, the contract exempted Cetus from sending all of the mutant strains they found, only that they would send improved strains. Among other flaws, it also stated that

Cetus would respect Schering's exclusive right to micromonospora, but said nothing about who owned the mutations that would naturally appear. No one noted the differing interpretations of "revenue generated"—to start, there was a world of difference between "net" and "gross" revenues that Schering would have to pay Cetus for their work. And no one thought to probe the legal definition of ownership of bioengineered organisms. The Schering contract, in short, was an empty agreement toward the principle of collaborative research, a messy first step toward the commercialization of bioengineering.[30]

For all its ambiguity, however, the Schering contract was also a watershed in business history. Instantly, it focused Cetus' energies on "strain-improvement," which made Glaser's machine a nominal piece of the company rather than a centerpiece. It also injected the company founders with much needed energy and restored the investors' confidence in them. Indeed, Cape and Alafi understood the significance of the moment when they intoned in a memo that "no commitment ever made by Cetus will be as important as that which we are presently undertaking [with] Schering," what with "the potential rewards so enormous." Finally, it defined bioengineering, at least for the moment, as a service for finding organisms rather than as a technique or technology used to make them.[31]

Building a Company and a Corporate Culture

From all sides, pressures played upon Cape and Farley to ready the company for its maiden contract by getting input from an expert in this or that bioscience subdiscipline. Not knowing whom to contact, or even which direction to turn, Cape decided to go much farther and abet the company's noble birth with the advice of as many elite scientists as possible. But the academics that Cape contacted, including Nobel laureates Arthur Kornberg and Paul Berg at Stanford and Gordon Tomkins at UCSF Medical Center, all clung to the professional maxim of separation of academia and industry, and showed no interest in helping a startup. Cape soon discovered that the mere mention of Don Glaser as a cofounder of Cetus would melt away ambivalences. Then came the decisive offer: Cetus would hire academic bioscientists as consultants and pay them generously—they would start with an offer of $2,000 in advance and $500 per day for twelve days of "work" each year, a sum that nicely subsidized typical academic salaries.[32]

Gradually, a few of the profession's elders signed up as consultants for the Cetus Scientific Advisory Board, and then urged the hoary canons of academic research. J. Yule Bogue, long considered the preeminent expert in pharmaceutical fermentation processes, was a professor of physiology at the University of London and had done a bit of consulting work for Imperial Chemical Industries in England in early 1960s. His advice to Cetus was spartan in its stark simplicity: "commit to long-term research budgeting—up to ten years." Some were long-time colleagues of Glaser at Berkeley, such as Henry Rapaport, and came more as a personal favor than for professional intrigue. The most committed academics to join the Cetus Scientific Advisory Board were Arnold "Artie" Demain, an applied microbiologist at MIT who specialized in vitamin and amino acid production, and Sir David Hopwood, a molecular microbiologist who studied antibiotic morphologies at the John Innes Centre and whose knighthood Cape always made sure to flaunt.[33]

Conspicuous among the scientific advisers was Joshua Lederberg, the geneticist from Stanford and a Nobel Prize winner, a consummate academic scientist and the most intellectually daring of the group. His leaping mind outpaced everyone else's—he was a modern-day Da Vinci—vaulting elegantly from deep analysis to sweeping conclusions to unintelligible rambling. At any given moment, Lederberg was totally committed to research in genetics, arms control and disarmament, pediatric birth defects, exobiology (the study of extraterrestrial life), biochemistry, something he called "cognitive biology," and so on. He also submitted the first grant request to the NIH to study a technique he called "gene stitching"—something his contemporaries would later call recombinant DNA. The peculiar thing about Lederberg was that he rarely stayed with an idea through its logical outcome. But this was consistent with the general haphazardness of Cetus too. A scientist interested in everything was a scientist naturally drawn to a company such as Cetus that had difficulty making up its collective mind.[34]

In all, Cape together with Glaser signed up about two dozen scientific advisers. It was an unprecedented collection of scientific talent and a novelty in American industry. Academic experts had played a role in industry before, particularly in industrial chemistry, but never so conspicuously, or so many with one company, or with so many noteworthy awards. They were newcomers to industry—professors, academics, and researchers from the ivory tower, idea men. The same facts that made them objects of intrigue within industry created an advisory board high on daring scientific input. That all of the SAB members were simultaneously presenting their scientific ideas as ideal research projects was an early indication of the wide-ranging, apparently indiscriminate eclecticism that marked Cetus' approach to running a bioengineering company.[35]

The Cetus SAB played no small role in guaranteeing that scientists would have preponderant influence in the company, but the final conversion came when Cape and Farley hired staff. In compressed time, between the Schering agreement in July 1973 and the contract's start date on 1 September, the two young executives interviewed everyone they could, with a preference for anyone trained in prestigious academic programs such as molecular biology at MIT, microbiology at Princeton, or biochemistry at Stanford. Profits may remain stubbornly elusive for four years or more, they said, but Cetus would always pay a competitive salary and assign central roles within. That was enough for Steven Goulden to come on board right away, from a temporary academic post in England to vice president of research, though he also played important roles in the antibiotic programs. Roy Merrill soon followed, holding down various responsibilities as director of computer facilities. Of course they hired specialists too, such as Bob Bruner, who supervised the assays department. Then a hierarchy took shape with the hiring of generalists such as Jay Groman, trained as an environmental biologist at the University Colorado, but who served as a research technician under Bruner in assays. Some, such as Jeffery Flatgaard and Beverly Wolf, filled no particular scientific need but could "do good science" in a variety of experimental fields. David Hansen, a talented physicist from Berkeley and an old friend of Cal Ward, accepted an appointment as director of engineering. Like everyone else, Hansen believed—and kept reassuring newcomers who had their doubts—that Cetus would use Glaser's machines to introduce bioengineering to entire industries. A few had no formal training in experimental biology but simply had a familiarity with the language and a marked ability to "learn as you go." And of course, they hired staff such as Douglas Miller to recruit and develop staff. Notably among the earliest Cetus employees was Terry Mahuron, who stood almost alone in the finance department as, simultaneously, controller, accountant, bookkeeper, and intermittently CFO.[36]

When Cetus hired its first employees, it was not at all obvious that they would identify with each other rather than their employers, or the company itself. The kind of informal hierarchy that Cape and Farley implemented looked as if it might become a rigid kind of corporate ladder that kept everyone dependent on their superiors. Scientists in particular were assigned specific responsibilities and reported to identifiable supervisors. These were general patterns, of course, because while Cape and Farley always appreciated their elite status at the top, neither they nor their partners showed any tendency to organize the company according to traditional patterns of authority.

The staff that Cape and Farley hired shared several remarkable characteristics, in addition to advanced academic training in a bioscience discipline. They came from all over, but the worlds they inhabited as graduate students—all attended school between the extremely formative years from 1961 to 1969—exposed them continually to ideas antagonistic to industrial capitalism and to higher education. Berkeley—both the city and the university—was simply an extension of what they already knew. They had spent years training for a life in academia and then made the fateful decision to become full-time employees of Cetus. In many ways, the company's first hires were a lot like the company's pioneers; everyone had taken on frightening risk and shared a sense that they were participating in a remarkable history. But in another way the employees had taken on greater risk because academia would not, at that time, allow anyone to return after they entered a commercial endeavor.

The new hires were enthusiastic about the novelty of their company, and that enthusiasm transferred into the working environment and social relations. For instance, business virtually shut down every time a Nobel laureate from the SAB gave a lunchtime presentation. After work the staff took great care to celebrate the birthdays of co-workers and to play on the company softball and volleyball teams, and on weekends they organized white-water rafting trips and went to see the Oakland A's, another highly successful team of nonconformists. The company founders may have imagined Cetus to be a great white whale—kind, well-liked, impressive, and cautious—but the employees printed t-shirts and buttons with a logo that looked more like the popular movie *Jaws*—dangerous and proud, and more than a little forbidding. That may have been the public image they preferred, but they made sure to meet each others' needs. They praised their co-workers' achievements in front of company executives; they shared special skills or services that others found useful, such as investment advice, a notary public, or legal aid, and there was always someone with medical training who could provide inoculations and tetanus shots on site. And when Bank of America rejected the loan request of one Cetus employee, the entire staff "moved in solidarity to negotiate a better [employee banking] deal with . . . Wells Fargo."[37]

To make the transition into a commercial industry more hospitable, new employees of Cetus drew extensively on the customs of the academic world they once inhabited to create, in a relatively short period of time, a unique corporate culture. Foremost, they understood that their livelihood depended upon the profitability of the company, so they took note of corporate revenues and protected intellectual property. But whether they conducted experiments, published articles in scholarly journals, or delivered papers at scientific conferences, Cetus employees continued to participate in a peer society that celebrated the most professional aspect of academic research. Bob Bruner recalls that the most challenging transition was the necessary deference to "the Bosses," because it made issues of authority and control more ambiguous than that which they experienced in academia. However, the influx of personnel put a premium on bench space and created close-knit quarters, says Jay Groman, which made it easy for peers to swap bacterial colonies, reagents, ideas, and craft lore: "everyone felt comfortable meddling in everyone else's work." This familiarity counterbalanced the authority of the company's executives, and then workers built common ground with organizing strategies designed to empower the scientific staff even further. For instance, the academic practice of organizing staff into formal working groups—a bio group, an engineering group, an assay group, a fermentation group—became a structure that supported self-management at Cetus.[38]

Scientific Promise and Company Peril

With a strong scientific base now in place, Cetus could finally sally forth on bioengineering "strain improvement" for the Schering contract. Everyone seemed to take great satisfaction in such a fruitful marriage of bioscience to industry, especially the company founders. Farley boasted that "we can carry out virtually any task or produce virtually any product. In short, anything that can be done, we can do better." Revealing a peculiar comfort with the market to defining the scope of the company, Cape intoned that "circumstances, not human will, has carried [Cetus] forward."[39]

But there would be no progress, only an occasional success followed by more problems and then great crisis. No one—from the founders, board of directors, scientific advisers, or company employees—suspected that the very confidence that had driven them to accomplish so much and had carried them so far had given birth to such a hurried, chaotic, and ultimately compromised company. Even the best-thought businesses take strange turns.

A quiver of foreboding crept into the company's vaunted status when one of Glaser's bioengineering machines broke down during an early screening run in late September 1973. It seemed innocent enough—the table upon which rested the petri dishes and laser scanner did not sit flat on the floor, which caused the bacterial colonies to shift and grow unevenly , making all the scanning results unreliable. The engineering group responded quickly, but then an employee trying to fix the table accidentally looked into a wayward laser beam and sustained an eye injury, so they shut down the entire system for a few days to build a makeshift cover. After the scanner was made safe to operate, somebody came up with the novel concept of using a block of wood to prop up the uneven table, but the scanner's drive mechanism moved back and forth with such force that the wood could not prevent the entire apparatus from moving around violently, leading some to worry about the "danger of crushing the user." So they bolted the table to the floor, but that caused the scanner to have "a chain and sprocket malfunction." The winter had been a wash, someone said, but the new year would bring better tidings.[40]

Instead, 1974 brought problems that swelled to horrifying proportions, well beyond what anyone could control. The first gaffe appeared suddenly, in late January, when the 100-liter sterilization tank began to leak because the engineering group had not tightened the bolts during assembly, which caused an entire batch of Schering's micromonospora to become contaminated. Sheepishly, they asked Schering for another batch and tried again. They resumed full-scale screening in early spring, but grave problems continued. No one thought to install a thermostat in the room that held the cultures, so the bioengineering group had "no way of monitoring the temperatures in the growing room," and there went another batch of micromonospora. Throughout all this, Cetus scientists somehow found a way to induce and identify improved mutation strains of gentymycin, and in the summer proudly sent a batch to Schering as proof that bioengineering could indeed work. Schering, however, claimed that Cetus abrogated the "good-faith" clause of the contract. According to Schering, Cetus should have delivered quantitative results, measured as a given number of dishes scanned in a given amount of time, a number of hours the scanning system was in continuous operation, or a number of strains improved. Schering's insistence pushed Cetus' scientists into crisis mode. They tried packing more cultures of micromonospora closer together on each individual petri dish, but that just contaminated the secreted gentymycin. Frustrated, the engineering group determined that they "needed a new system" to produce gentymycin strains faster and began the rather elaborate process of reengineering the entire lab. After months of trial and error, with the new production system almost three-quarters of the way complete, "all hell broke loose" when a critical member of the engineering group suffered an "untimely juxtaposition of a bicycle tire with a drainage grating, causing an impact between the rider and the roadway."[41]

The hell that seemed to follow Cetus throughout 1973 and early 1974 might have been dismissed as low comedy except for the uncomfortable fact that none of the company leaders could be found. Alafi had already begun exploratory work for an initial public offering, which took him far away from daily business routines. Glaser, the one individual who knew the most about the machines, continued to face down a frenzied attack by his colleagues at Berkeley, many of whom now openly criticized commercial ties in academia as "an ethical 'deep structural' poverty of scientism [*sic*]." Cape desperately wanted to be the boss that everyone liked, except that he enjoyed hobnobbing with the clientele even more, accepting invitations to speak about the future of bioengineering at college campuses across the country and crisscrossing the globe in search of new international divisions for the company. And Farley was so busy flitting about dinner dates, theater engagements, and Playboy Clubs that no one could get in to see him.[42]

Then, 7 March 1974, a fire broke out at the newest Cetus facility on Fourth Street, causing extensive damage to expensive experimental equipment and the destruction of yet another batch of Schering's micromonospora.[43]

Unsure what to do or even which direction the business should go, Cape made the reflexive decision to call the company's scientific advisory board for more advice. Everyone referred him to Bill Bogue to right the company. His decades of experience with industrial fermentation and his unwavering faith in "old-school scientific methods" made him seem like the ideal scientist to identify the source of the problem and a possible solution. In him the rest of the scientific advisory board saw a uniquely perceptive observer who could be counted on to speak with candor, insight, and moxie.

Bogue did not merely report about Cetus. His "unfavourable" rebuke broke the company's prevailing optimism. Said Bogue, cogently:[44]

> The operation as at present constituted is, in effect, an enlarged academic facility rather than an adapted one. Adaptation to industrial requirements demands a different attitude towards housekeeping, stricter discipline, stricter routine monitoring, improved barriers to cross contamination and to stray contaminants and a foolproof flow pattern. . . . I do not recollect seeing any room or work area with really good housekeeping. In those areas in which several people were working, I had the impression of . . . chaos.

Bogue lambasted the lack of cleanliness and listed the gross negligence that he had seen:

- Experiments [were] conducted on bare wood strips and [near] tile joints in sterile areas that were rough and probably absorbent
- It is not very impressive to observe an individual . . . wearing protective gloves to . . . handle the telephone, clipboards, etc.
- Laboratory footware is really essential
- The flow pattern of materials to and from wash areas allowed clean and dirty materials to cross—not a good idea
- Clothing: I was worried by seeing personnel moving around in their street clothes and outdoor shoes
- In the "clean" corridor I noted two pairs of used disposable protective pants thrown over a carton containing a new supply of them.

Inexplicably, on a return visit Bogue saw more of the same:[45]

- Attitudes toward mutagens should be similar to that toward most other dangerous chemicals, such as ether or cyanide.
- There are general precautionary procedures which one always follows when using dangerous chemicals, such as not pipetting them by mouth.

On the whole, it was a bad report for the scientists at Cetus, which was ironic. Nothing meant more to the company, no one had done more work, and in terms of sheer numbers, there were simply too many of them to dismiss their contributions so heedlessly. Furthermore, conspicuously absent among Bogue's reports was any reference to the company's leadership. A relieved Farley reported back to Bogue that his report had compelled Cetus to implement a series of new policies: "Thursdays and Fridays are the days requiring the highest level of sterility." He said nothing about Cetus' sterilization policies on Mondays, Tuesdays, or Wednesdays.[46]

In the months following Bogue's report, the mood at Cetus turned gloomy. Then matters came to a head when Alafi decided at last to resurface. True to his instincts and consistent with Bogue's overall assessment, Alafi lambasted the scientists for their total disregard for the financial health of the company. To Alafi, it was not merely that the machines kept breaking, or that Schering kept reneging on the contract, but that no one seemed to care about the company's prosperity, or the concerns of the investors behind the scenes. This was only half-true. The company had indeed burned an extraordinary amount of cash in a relatively short period of time, but the scientists could hardly be held accountable for their own swelling ranks. Most of all, however, on the issue of cash flow, the different perspective of the venture capitalist became a genuine antagonism.

"We need 8 scientists . . . for a 7 million/lab/year?" wrote Alafi in a barrage of sarcastic memos. Feeling betrayed, the scientists united and formed a "Safety Committee," and firing off a general memo that stated in no uncertain terms that all of the company's failures "can easily be propagated to upper management." Never one to back down, Alafi volleyed back, "the Safety Committee costs $75/hour, or about $100 each meeting," and ordered Cape to break it up, which he did to great discomfort. Alafi's criticism of the scientists inspired Terry Mahuron, the lone financial adviser at Cetus, who raged against the scientists for their "total disregard" for money and resources. The charges turned out to be baseless, but Alafi thought he smelled a scandal anyway. He invited a statistician from Berkeley to come behind closed doors and study the accuracy rate of the bioengineering system implemented by Cetus. In terms of pure probability, said Alafi's mole, "you could flip a coin and you would do better." Wasting no time Alafi angrily confronted Glaser and accused him of overstating the capability of bioengineering and misleading him into a reckless business venture. For the first time that anyone at Cetus could recall, Glaser spoke with forceful reassurance to Alafi: "first you walk, then you stumble, then you run. Science proceeds just like this, and so will our science and machine. But it will work." It also happened to be the first time that anyone at Cetus had counseled patience.[47]

One person was not displeased with the turn of events at Cetus. That was Pete Farley, the wild card in the deck of company leaders, determined to play the salesman's hand that would save the company from certain financial ruin. Never mind that the scientists could not fulfill just one contract when Farley knew that Cetus had the opportunity to take bioengineering everywhere. He tried the improbable task of convincing Stanford University to enter into a commercial relationship "wherein Cetus had the licensing rights to any commercial application of their projects." On the grounds that Stanford was not a private company and would not consider giving away its intellectual property anyway, university administration decided to pass on Farley's proposals. He approached nationalized pharmaceutical manufacturers in India and the Philippines and offered bids to "screen huge quantities of antibiotics," but they said no too. Undaunted, he resurrected the first market that Cetus considered—System A, or the manufacturing and sale of the microbial screeners. He could almost see a cheaper version of Glaser's bioengineering machines selling for "$600—$1,000 each, depending upon the number of knobs and whistles and sex appeal . . . and the overall parameters of the market." Manufacturing offered additional benefits, added Farley, because it created new markets for products compatible only to the Cetus system. "Square petri dishes" was just one of many possibilities, and he hastened to assure that the market was enormous too: "square petri dishes w/agar at $50,000,000.00/yr, and w/o agar at $20,000,000.00, along with an assortment of filter paper $25,000,000.00 and report forms $5,000,000.00." Farley could not believe that "the financial community and all the other idiots in the world who get very riled up about large recurring markets have missed this opportunity. I can't, off the top of my head, think of a good reason why everybody in the world wouldn't want one." In Farley, just as it had with other executives at Cetus, business trumped science again—the size of the petri-dish market was, and remains, roughly the same size as the market for square wheels.[48]

Nevertheless, rejection merely galvanized Farley. Sure enough, in winter 1974, Farley closed deals with Upjohn to screen for Erythromycin strains and a general research contract with Bayer, with the upside that both companies would pay some of their fees up front. The scientists were beside themselves, but Alafi defended both contracts and even encouraged Farley to negotiate with industrial giants such as Glaxo, Delft, Ciba-Geigy, Stauffer, and others because he wanted accountable evidence of "earned income" to prop up his $5-million financing deal already underway.[49]

Having been pushed into precisely the kind of situation they most wanted to avoid, the scientists now found themselves trying to cram three bioengineering projects into one. Theoretically, Cetus needed to examine 250,000 sets of organisms each quarter just to meet Schering's minimum expectation, to say nothing about the new quotas for Upjohn and Bayer. Unfortunately, after one full quarter under pressed conditions, they completed 80,000 screenings, which meant they needed to do about 420,000 in the next quarter, or about 192,000 each month, just to fulfill their contractual obligation to Schering. To catch up, they hired six new scientists and ten technicians, and they still needed more, so they added an additional six scientists and dozens of technicians. Saddled with unending maintenance problems and impossible production schedules that they could not possibly expect to meet, a few employees simply walked away. Cetus had little choice but to halt the Upjohn and Bayer projects, which created more legal problems, and forced them to concentrate entirely on their original contract with Schering.[50]

Someone had to take control of the company. The pivotal figure in this group was no longer Glaser or Alafi, and it certainly was not Farley. The bioengineering group was back on their heels and hesitated to take charge too. So, by default, the moment belonged to Ron Cape, the man who, almost literally, lived for conversation and notoriety. He had a delicate command of the science, but a deep respect for those who did, and so was more comfortable following the advice of the SAB than guiding them. However, he had an attractive personality and always a story or a quip that could disarm conflict. He may not have been the ideal person to save the company, but at the very least he could always calm jittery nerves. If nothing else, Cetus sorely needed this.

Throughout spring and summer 1974, Cape contacted virtually every consultant from the company's SAB and confessed that Cetus had trouble hitting Schering's moving target, contractual and otherwise. Many of those contacted scoffed when they heard that the company had focused its fundamental efforts so narrowly on bioengineering and called for "immediate and maximum diversification" of scientific research into as many commercial fields as possible: "a. stick to [original] antibiotic projects, b. respond to random stimuli like the Israeli oil-spill, c. aggressively . . . go after large-scale opportunities: chemical/fermentation plant design, waste water engineering, mining and oil extraction, etc."[51]

Cape embraced the SAB's entire package of proposals with characteristic enthusiasm, which set off a wild "market-driven" approach to research that spread resources and personnel across a wide spectrum of projects. With an eye toward solving the energy crisis, Cetus recruited scientific advisers skilled in "continuous cellulase production," such as the chemical engineer Dr. Charles Wilke, and shifted Jon Raymond and his screening research team out of the Schering project and into development work on "cellulose . . . to harness the solar energy stored in plants." They took suggestions for treating chitin (crab shells) to make thin transparent film used in food packaging, then expected heroic returns: "everyone wins—seafood processors, environmentalists, government, scientists, industry, consumers, and the ocean." Almost casually, Cetus tacked on to the gentymycin project an identical strain-improvement program for cephalosporin antibiotics. Here again, Cetus attempted to capitalize on a unique market opportunity: Eli Lilly had earned about $325 million in annual revenues in that market alone, but stood to lose its patent protection at the close of the decade. Some of the new projects seemed impulsive, such as the experiments with chenodesoxycholic, an acid used by physicians to dissolve gallstones. However, the decision to collaborate with Schering Agriculture on a "steroid conversion" project, despite on-going problems with the parent company and the technical limits of the screening system to accurately identify anything but bacterial colonies, simply defies any rational explanation. These, to say nothing of the requisite projects on citric acids, ethanol, sisomicin, vitamin B-12, and so on, pushed bioengineering aside. Such diversity strained the financial resources and the staff, which was neither stable enough nor large enough to perform this array of tasks. Yet, rather than oppose such a shift in strategy, the executive board likened the move to "financial diversification" and approved it, while the scientific staff seemed genuinely appreciative of the professional autonomy that went hand-in-hand with open-ended research.[52]

The scientific advisory board at Cetus was not the only observer searching for commercial opportunities in the biological sciences. A young upstart venture capitalist who had come from the Kleiner-Perkins firm, a second-round investor in Cetus, twenty-seven-year-old Robert Swanson, also saw reason to explore bioengineering. At some time in early summer 1974, Swanson told the two partners in his firm about a celebrated article in the *New York Times*, "Animal Gene Shifted to Bacteria; Aid Seen to Medicine and Farm." As far as Swanson could tell, new recombinant DNA techniques developed by Paul Berg and Stanley Cohen of Stanford University and Herbert Boyer at UCSF could be used to bioengineer—or, literally *make*—proteins that met "some of the most fundamental needs of both medicine and agriculture." On behalf of Swanson, the firm's partner Tom Perkins placed a call to Alafi and asked if he and his upstart associate could reconnoiter the world's most promising bioengineering company. Alafi proudly led them on a tour of Cetus, showing off the many different projects, their market projections, the talents of the staff, the different facilities, and eventually Glaser's bioengineering system. Sufficiently impressed, Swanson issued a bold proposal: he would manage a second bioengineering division for Cetus, a recombinant DNA program, which would sit alongside the bioengineering system already in progress.[53]

Cetus could hardly ignore Swanson's offer. One year earlier, Hoffmann-LaRoche had approached Cetus with a request to explore the "possibility of using recombination," but at the time, Cetus was wholly committed to using Glaser's machines as a bioengineering system and turned them down. At about the same time, executives from GE mentioned they had a "modest but effective [genetic engineering] procedure used by Al Chakrabarty and Steve Rosenburg," but again Cetus did not pursue this offer. Cetus could credibly believe that their own bioengineering program, radical enough by any objective standard, was prudent and attainable when compared against the embryonic techniques of recombinant DNA. But Swanson's offer was different. It was direct, it would occur within the walls of their own company, and it had the backing of willing and familiar investors.[54]

Ignorance had enshrouded Cetus leadership. No one at Cetus—including Glaser, whose scientific insights were keener than most—anticipated the scientific or commercial merits of recombinant DNA. Moreover, the founders were not only unable to remain focused on the industry they longed to create; they were almost equally unable to see how a company could use recombinant DNA as a commercial venture. Under these circumstances, it was not simply the weakened state of the company but reasonable skepticism that led Cape to their mercurial scientific adviser, Joshua Lederberg.

Fatefully, Lederberg expected there were "other ways to make more valuable products" than just recombinant DNA—a veritable "no" considering there never was another scientific project that he did not like.[55]

History will not look kindly upon Lederberg's advice, or Cape's decision to approach him, but at the time there was no one better to ask. Lederberg was the first to propose gene-stitching to the NIH in 1967. In 1971, he served as the faculty adviser on two different recombinant DNA projects when few, if any, had students practicing in the field. His office and laboratory neighbored Stanley Cohen and Paul Berg's in the Stanford University Medical Center, arguably the point of origin for the first successful recombinant DNA experiments. Further, given his academic position and his experiences with private companies such as Cetus, he understood as well as anyone the slow pace of research and development in academia and in industry. In short, Lederberg's opinions represented the purest, most informed scientific orthodoxy available.

This much is also clear: the consequences of Lederberg's lukewarm endorsement were immense. Cetus would follow his advice and stay its current course, using Glaser's bioengineering machines to pursue a wild plurality of commercial opportunities with a sometimes desperate fervor. Most would fade from view as unfulfilled alchemic promises, while recombinant DNA would challenge Glaser's microbial screening system as the scientific platform for bioengineering. And, from what was still in 1974 a single biotechnology company would spin an unprecedented industry—unrestrained.

Down through Swanson's offer to run a recombinant DNA division, Cetus seemed by all appearances poised and ready to become a wildly successful company. The press lauded Cetus as a "progressive" company with an abundance of scientific answers to society's most pressing needs, and compared it on equal footing with such powerhouses as IBM, Intel, Hewlett Packard, and Microsoft. Investors widely believed that Cetus would deliver on its promise and clamored for an opportunity to invest. The sheer scale of the scientific activity of this new company no doubt helped to shore up its public image, as did Cape and Farley's own private proclamations, "we will essentially ransom the world!" Despite the exhilaration of the public, despite the assertions of its founders, despite the exertions of its scientists, despite all the ingenuity and exuberance, one key truth stands out: something had gone terribly wrong at Cetus.[56]

Why did Cetus struggle when it had so many advantages? Answers converge from a number of directions. First, in terms of development, Glaser's bioengineering machines clearly failed as an operable tool at industrial production levels. From this perspective, Glaser's system sounds like a straightforward mechanical failure and should never have been attempted. Simply put, was it a mistake for the founders to use Glaser's screening system as the technology in which to launch a commercial biotechnology company?

Certainly not, at least from the perspective of the participants. The social, scientific, and financial rewards for successfully developing a continuous process for bioengineering are undeniably enormous, then or now. Even if the project took many years and much money, everyone believed it should be pursued until the system was proven successful or impossible.

If errors of judgment were committed, they occurred in the tone of the venture and the pace in which it was executed. The tone of the company's operations was set by the scientists. No one at the time thought of this as a disadvantage. In a sense, the founders considered scientific excellence a necessary advantage and the basis for their existence as a company. Metaphorically, Glaser was used as the symbol to represent the scientific merit that defined the company: they had him sign virtually every piece of correspondence that went out, they inserted his name into as many conversations as possible, and despite his overwhelming reluctance, they included him whenever they went before the board or the company's investors. In turn, whenever the founders thought they saw the company unravel, they turned decision-making authority over to the SAB. Moreover, three of the founding partners, the first and then almost all of the company's staff, and key members of the board of directors all had advanced training in a life-science discipline. With such a strong scientific base, the venture capitalists who knew little about the biological sciences uncritically deferred to its practitioners and almost always accepted the data presented by the scientists over their own best judgment. "Investors like Ed Carter Hale always seemed in awe of a Ph.D.," said Cape, who had one. Undeniably, everyone approached Glaser with the same reverence and awe; intoxicated by their proximity to Glaser, they all wanted to believe that the laureate could accomplish anything. This respect for science manifests itself, at various times, as adversarial, respectful, and independent. Indeed, there were occasions when Cetus needed help from the Berkeley faculty, or when its own failing projects were being duplicated in Donald Glaser's academic laboratory on the Berkeley campus. Yet, Cetus leadership decided to keep all of the company's research and development at great intellectual and physical distance from research and development at the university. Their stubborn insistence on doing everything internally would be one of the defining traits that separated Cetus from all other biotech companies to follow.[57]

While scientists set the tone, the pace of the company was set by the venture capitalists. The ease with which venture capital was willing to overlook business details attests to their hope and their willingness to give money to scientific experts, especially a company led by a Nobel laureate. Their ambitions overshadowed the discipline necessary to perform due diligence on their investment, further fueling the hysteria. Everyone was champing at the bit to get started. They felt the pull of potential profits and were pushed by the fear that competition loomed on the horizon. Leading the charge was Moshe Alafi, whose drive for the fastest route to the biggest returns easily captivated the younger nonventure capitalists such as Cape, Farley, and Ward to get caught up. Furthermore, the influence of venture capitalists as investors and board members—of interest, money, and influence—contained serious deficiencies as well. They were revolutionaries, they imagined, on a mission to build a new industry with Glaser's machine. What's more, they believed, they were fated to succeed. It seemed never to have occurred to any of them that the machine might not be ready, or that another system might replace it.

In the end, power within the company vacillated between the scientists, who understood bioengineering the best, and the venture capitalists, whose reign over commerce the scientists could not contest. Together, they ran at a youthful and impatient pace. Internecine warfare cancelled out the individual talents of the two sides, a happy by-product for managers like Cape and Farley, who found tremendous concentrated authority. Nevertheless, despite blatant mismanagement, Cetus accomplished one major feat—they had vanquished the stubborn habits of "old friends." With that, there was just one last obstacle to overcome before a bioengineering industry could proceed unrestrained: the magnitude of popular concern that previous generations of bioscientists had provoked.

Conclusion

An End . . .

Science is changing. It is an insidious change. It's now a complicated business; everyone wants mission-oriented science. . . . The attitude . . . is that all of society's problems can be solved in the next decade—should be solved—these are changing expectations and no one, it seems is patiently interested in nature anymore.

—Sydney Brenner

The savage collision of all the forces that had swirled around Bay Area bioscience programs since World War II—precarious interdisciplinary tension, a disruptive political culture, and an unraveling political economy, not to mention the dramatic surge of applied bioscience discoveries in recent years—occurred on 20 February 1975. On this date, investigators from around the world came to the Asilomar Conference Center in Monterey's Pacific Grove, to find common purpose, to replace social and fiscal tumult with a more stable footing, and to determine nothing less than the outermost limits of applied bioscience research: genetic engineering. They came expecting Asilomar to be "the pivotal event" in bioscience history, at least since the discovery of the structure of DNA.[1]

But Asilomar would produce few good solutions, only precarious agreements followed by further confusion and then even greater crisis. However, history takes unexpected pivots. Indeed, it was not Asilomar but private interests that provided a startlingly clear solution to the federal government's unrelenting cutbacks in funding and the ceaseless frustrations that the public had toward the biosciences. In moving closer to the mainstream of capitalist interests, Bay Area bioscientists, even without fully intending to do so, protected an almost wholly new field, the biotechnology industry. In the final analysis, their precipitated decision to accommodate capital brought an abrupt end to the confrontations between tradition and revolution, and between science and society, that had plagued the discipline since World War II.

Bioscientists came to Asilomar primarily because Paul Berg asked them to in a letter published in the July 1974 issue of *Science,* which also called for a voluntary moratorium on recombinant DNA research until risks could be assessed and research guidelines could be established. Berg called the conference not because he opposed genetic engineering, but because he found other developments peripheral to it deeply troubling: the "quickening pace of scientific research," the lack of respect for "the sanctity of bioscience traditions," and the very real possibility that a recent bioscience discovery—cloning, for instance—might unleash an "uncontrollable biohazard."[2]

Many of Berg's colleagues who had also come of age during the basic bioscience heyday shared his concerns. Certainly the spectacle of surging applied research would not end their quest for fundamental knowledge; no revolution could. Nevertheless, a good many investigators found it a little bewildering that a stable field could be so consumed by practical concerns. And they looked to Asilomar as their last good chance to regenerate tradition—as a means of reviewing, reestablishing, and purging this new field of its extreme qualities. "The pause," said one investigator, "will prompt workers in this field to assess the situation." Another cautioned that "to ignore the potentiality of [cloning] to wreak major imbalance in natural biochemical cycles seems like straining a gnat and swallowing a camel." Sometimes an opinion seemed more a righteous sense than anything concrete in the here and now: "I think a [meeting at Asilomar] may be a good idea." Most fixed on biohazards as the extreme result of the sorry mess that applied research had caused. All other concerns shriveled to trivial proportions in comparison: "I hope [Asilomar] will address accidental hazards of 'shot gun' type experiments in a specific way. It cannot however ignore the second category of . . . purposeful hazards."[3]

For those conducting genetic engineering experiments, Asilomar provoked gloom. The temporary moratorium imposed on their research challenged the profession's foremost birth right: independence and autonomy. They used words like "defeatist," "irresponsible," and "threatening to the future of research" to describe limits imposed upon their work and demanded a return to open experimentation. "Rather than raise the specter of moratoria of any sort," countered one of Berg's many opponents, "let each scientists decide at the outset of his experiments whether he would care to expose himself, or better, his child, to the newly assembled 'agent'. . . . I've little doubt that . . . any damage done to our species by careless or heedless researchers would be trivial in comparison with the seriousness of the loss of freedom of inquiry."[4]

Callous though this position may have been, the offensive they waged against the biohazard argument was an ingenious full-scale defense as health saviors, a bold front to growing genetic diseases and starvation. Sometimes their enthusiasm went unchecked: genetic engineering had no limits; it could, as one proponent fantasized, "improve education, reduce misery, minimize personal and group conflict, prevent and treat mental illness, and relieve parents of guilt based on exaggerated assumptions about the range of their influence on behavior." Even economists such as Milton Friedman celebrated the search for practical benefits through genetic engineering as the best motivation for innovation that a society could have.[5]

Behind the high-flying rhetoric and the progressive idioms that cast genetic engineering as having transcendental social purpose, most investigators appreciated the benefits that this new frontier bestowed. The typical bioscientist lurking in the academic or industrial laboratory hallway was young, unestablished, saddled with debt, and untethered to any single university. Many had survived the initial contraction of federal patronage by applying fundamental knowledge to practical concerns through genetic engineering; on the eve of Asilomar, the NIH alone funded thousands of experiments that involved either the insertion of foreign DNA into mammalian cells, or cloning, or both. The availability of federal money for genetic engineering had a way of expanding the field's scientific interests while at the same time attenuating them.[6]

Most of the public thought about recombinant DNA experimentation with detached indifference. It was easy for people to be scornful: few understood the science and it was easy to see that a significant amount of tax dollars supported all kinds of research, which muted scientists' protest and made their cries of poverty shallow. And for more than a decade, a forceful counter culture had railed against scientific abuse, which pushed popular sympathies instinctively toward those who wanted to prevent scientific catastrophe rather than toward those who predicted scientific salvation. In short, most of the public appreciated Berg's plea for caution.

Some people were hysterical about genetic engineering and the possible threats that it posed to society, but their passion did not make their arguments any more convincing, or more right. Many of the nostrums that resisted this research tested the limits of rationality. One woman wrote a key participant in the first cloning experiment at UCSF to ask if it was "possible that she had been cloned." She remembered being abducted by a spaceship, then she blacked out until she was returned. She was convinced that the space people had cloned her: "Is there something I could look for, like a seam?" Others lived in constant fear of a new "dreadful virus," and cowed by insecurity, offered their "appreciation for any effort, . . . anything that can be done to stop [genetic engineering]."[7]

More conventional public opinion could hardly be called rational either. Many disaffected conservatives found all government-sponsored research threatening, a populist concern that bordered on paranoia. *The Economist*, usually a thoughtful periodical, nevertheless pointed an accusatory finger at any bioscientist who refused to acknowledge that genetic engineering could cause a "catastrophic biohazard." Other conservatives who wanted to restrict genetic engineering did so for spiritual reasons, though it would have been hard to call most of this group devout in any other setting. Even those whose opinions were reasonably based on faith became energetically opposed to this kind of research, not out of religious devotion, but because the open and universal agnosticism of bioscientists provoked them. Yet, there were also a significant number of conservatives who spoke stubbornly in support of genetic engineering as a right of "competition," "individual initiative," and "free enterprise." Of course, this argument rested on traditional capitalist avenues of private property and libertarian conceptions of unrestrained freedom rather than science and concern for public health.[8]

The deepest public opposition to genetic engineering came from the left, as committed environmentalists such as Jeremy Rifkin and Theodore Roszac believed this type of experiment threatened the natural evolutionary order. For more than a decade, they fought a valiant war to protect the "natural" against harmful pharmaceuticals, unsafe gas pipelines, food additives, tainted meat, pollution, herbicides, and radiation emissions. On the eve of Asilomar, however, the debate about biohazards became paramount. But the environmental argument was inherently weak too, because they did not have, as yet, direct evidence to defend their position. Moreover, the conservative nature of their argument—in times of disorderly pursuit of progress, they believed in returning to the natural system—failed to connect with a majority, which made it difficult for anyone in the public sphere to speak with a unified voice against the new bioscience order. And of course, blanket opposition to genetic engineering for any reason was not entirely fair. Scientifically, genetic recombination can occur naturally. Moreover, few humanitarians, liberals, conservatives, environmentalists, Christians, or populists could categorically oppose synthetically bioengineered microbes that can eat up oil spills, bioengineered foods that can feed the starving masses, or that perpetual dream, bioengineered pharmaceuticals that can cure disease like cancer. The painful fact was that however loud the public jeered genetic engineering, however often they said that they were opposed to it, in this confusing world and on this technically complex topic, they were not really sure what to think, or how.[9]

Bioscientists' accumulating grievances, compounded by the public's desire to engage in the debate despite a profound lack of understanding, made the days before the Asilomar Conference particularly ripe for incendiary rhetoric. All sides dug in, and everyone it seemed turned on Paul Berg as their unwitting chief adversary. Those who opposed the moratorium on genetic engineering howled: "You . . . are unduly alarmist"; "the statement that an accident could kill millions and that a virus could be constructed that would kill everyone is absurd"; and "if there is no documented evidence of danger, proceed until you suffer the consequences of that danger." Not to be outdone, those who supported the moratorium categorically demanded: "if there is the remotest chance of trouble, don't do [the research]." Apparently, neither side saw irony in the charges put forth by the scientists that the public was behaving hysterically. Privately, Berg knew there was a certain inevitability to genetic engineering—someone somewhere was bound to conduct the research. In a desperate attempt to tread water in a sea of tumult, Berg wrote to many of his critics and qualified his position: "[I] never called for a ban on genetic research; nor . . . objected to this line of investigation on moral or ethical grounds, i.e., because it would be opening the door to God knows what!" "Keeping things neatly in-house," Berg opined, superseded all other concerns.[10]

Since all bioscientists could agree that an open debate about genetic engineering posed a greater threat than biohazards, Berg and other conference organizers made the fateful decision to pursue Joshua Lederberg's suggestion to "control the seeds of discord." To reduce the possibility of mischief, organizers carefully invited 150 people to the Asilomar Conference, most of whom were intimately involved in genetic engineering, but some of whom came from industry or had an affiliation with the military. In contrast, no investigator who opposed genetic engineering could attend and only one social activist—a spokesperson for Science for the People—was invited, but they made no attempt to replace him when he replied that he could not attend. The prefigured agenda squelched all discussion of compliance, regulation, and enforcement, and only one evening presentation and one lunch-time forum addressed ethical considerations. Furthermore, organizers limited press coverage to eight reporters, the taping of any meeting was restricted, and no reporter could file a story until after the end of the meeting; only the threat of a lawsuit by the *Washington Post* and the ACLU forced conference organizers to capitulate, but even then only thirteen more reporters were invited to cover the event—with limited access of course.[11]

The careful preparations and the sanitized agenda helped turn the Asilomar Conference into a carefully staged defense of genetic engineering, interrupted by occasional attacks on a few high-profile experiments. As planned, virtually all of the presentations centered on technical aspects of recombinant DNA research or biohazards that occur in laboratories. The introductory and closing remarks served as memorable bookends to what was a doggedly science-centered and scientist-controlled conference on genetic engineering. Berg's cochair, David Baltimore, opened the proceedings by cautioning everyone on what could and could not be discussed over the course of the next three days:

> This meeting was conceived to lay out the existing technology, to consider what has been done, what might be done . . . and what benefits can come from [genetic engineering] both in terms of knowledge and in more practical terms. But the impetus to call the conference was one of concern about potential hazards and about the safety of the techniques. So the ultimate focus of discussion . . . must rule out topics peripheral to this meeting . . . : the complicated question of what's right and what's wrong, complicated questions of political motivation, . . . and the potentiality to utilize this technology for biological warfare.

The conference concluded with what one eyewitness described as a "sobering . . . and discomforting" presentation by two corporate attorneys who offered a poignant reminder that the greatest threat to genetic engineering came not from within the discipline, but from the public: "Many have talked about their research under the banner of 'academic freedom.' By overstating their case they risk provoking greater restriction. Freedom of thought does not encompass freedom to cause physical injury to others. . . . It is absolutely an appropriate response [of the public to demand restrictions] where irreversible harm is threatened." "The public has every right," cautioned the other attorney, "to get involved in the debate."[12]

Naturally, suspicions about Asilomar ran deep within public circles. The most outspoken critics then and later insisted that the conference provided a lightning demonstration that bioscientists aggrandize their own authority by eliminating debate. A groundswell of reporters railed that bioscientists genuinely did not care about biohazards and held contempt for democracy. Science for the People circulated a letter among those who attended the conference demanding open participation at all future events: "We see in the structure of this conference that a scientific elite is . . . trying to determine the direction that such regulation should take. . . . We do not believe [this group] is capable of wisely regulating this development alone. This is like asking the tobacco industry to limit the manufacture of cigarettes." One scientist who had not been invited to the proceedings "slapped the Bishops of Asilomar" with a scathing article for *Science* that resurrected as a metaphor certain repressive symbols of medieval society: "At this Council of Asilomar there congregated the molecular bishops and church fathers from all over the world, in order to condemn the heresies of which they themselves had been the first and the principal perpetrators. This was probably the first time in history that the incendiaries formed their own fire brigade. Their edict, . . . which lists various forbidden items, reads like a combined curriculum vitae of the conveners of the conference."[13]

In the end, the Asilomar Conference achieved only one of its objectives: it had cosseted genetic engineering from public debate. In a perverse way, the public's desire to intervene had rallied investigators around the one issue that gave the discipline its keenest edge: autonomy. It was on this issue, and this issue alone, that all bioscientists could always agree. And Asilomar convinced enough of the public that investigators would watch, monitor, and regulate their own experiments. By making public that they generally supported and had control over genetic engineering, they legitimized this new direction and solidified their authority.

In terms of policy, however, Asilomar provided little direction. A few reforms were passed, but these scarcely ended disciplinary division, nor even the debate over genetic engineering, especially concerns about biohazards. Investigators generally agreed to temper their own individualist, ambitious tendencies, and they would conduct thorough reviews of their own laboratory practices. They also agreed that a select group of representatives would codify cautionary sentiment into a formal policy designed to tame the roughest edges of genetic engineering. But without a formal oversight provision, even these policies went virtually unenforced. Simply put, investigators were responsible for policing themselves. In terms of policy, therefore, Asilomar produced a stalemate no less intractable than the stalemate that had paralyzed collaboration between pure and applied research in the decades before. Indeed, at many points it seemed that each side was less the principled opponent than its own willing captive. Those eager to get on with genetic engineering sincerely believed that the conference had constrained their professional opportunities and rendered them effectively powerless in the face of the financial crisis that continued to unfold. Traditionalists blustered against the new policy, calling it a sell out and a "watered-down version" of the intent of the conference. In the end, bioscientists could not see the future of genetic engineering very clearly. It was as if the entire field, poised to enter a phase of experimentation more dynamic than ever before in its history, felt the need to reassess the relative importance of pure discovery before rearranging all research priorities yet again.[14]

. . . And a Beginning

After Asilomar, the intractable difficulties in the domain of popular unrest and shrinking federal patronage continued. The nationwide recession in the early 1970s, the worst economic conditions since the Depression, exacerbated the calamitous cycle that had begun in the mid-1960s, while federal money that remained typically went to research projects not conducted in California. Furthermore, Bay Area bioscience programs confronted an especially distrustful public who continued to look upon the field with suspicion. Who could have predicted that a region that had grown from a mere afterthought into an academic leader in a single generation could become so fragile in the next? That the people had finally secured a new kind of practical bioscience research would continue to look upon it with great reluctance? Or that the Congress that had deferred to the authority of the bioscientists and that had provided seemingly unlimited research patronage for almost two decades would have the audacity to ask for greater research utility without also offering sufficient support? The fortunes of the field may have been a surprise by 1975, but an enduring fact had become painfully clear: the tide of the biosciences had ebbed relentlessly.

The fate of bioscience research in the Bay Area turned in particular on the issue of finding research support, and although federal patronage had not disappeared, its decline was bad enough to greatly complicate everyone's experiments. Entire programs faced the prospect of decay. UCSF's accounting office gave full vent to the crisis, and with subtle threat, distributed to all staff involved in research "nasty memos on their desperate fiscal situation." "You are spending $20,000/month," fumed a university budget officer to the chairman of the biochemistry department: "This means you have less than 4 months of money left. . . . It also means that you are more than 3 month short, or $60,000. . . . We HAVE to come up with $60,000 in the next couple of months—I don't know where the money will come from if you don't. PLEASE GIVE THIS SOME ATTENTION!" Throughout the postwar period, Berkeley never wavered from its steadfast pursuit of a top-flight bioscience program, but the financial problems in the mid-1970s were so bad that it forced the university to postpone its perpetual dream: "the biological sciences cannot develop beyond their present 'primitive' level," declared a demoralized administrator, "because funds are no longer available." An official at Stanford boldly pronounced that in order to withstand the budget onslaught, research staff in the medical school should consider all options as reasonable: "It is a fact that the financing of . . . universities is more difficult now than at any time in recent memory and that the most likely prediction for the future is that a hard struggle will be required to maintain their quality. . . . To put the point as precisely as I can, we cannot lightly discard the possibility of significant income that is derived from activity that is legal, ethical, and not destructive of the values of the institution." This Stanford official saw clearly what the rest of the field wanted to avoid: that support would come from less-than-ideal funding sources, and then everyone would have to attenuate their values accordingly. Clearly, and for the first time in recent memory, investigators at all three research universities stood at the door of a forbidding experimental threshold, unsure if they should enter a new chapter in the history of bioscience research.[15]

The common financial struggle assuredly did not evoke public sympathy, or agreement among bioscientists about its remedy. Over the raucous objections of traditionalists committed to pure research, many local investigators began turning to private industry as an alternate source of patronage. Donald Glaser, the physicist-turned-bioscientist who strong-armed Berkeley into forming a molecular biology department, co-founded Cetus Corporation, one of the first biological research firms. Joshua Lederberg, the Stanford geneticist who had become stifled by the lack of opportunity in academia, accepted an offer to serve on Cetus' board. So many investigators in Stanford's medical school had obtained a patent for their work that the royalties they earned made up 35 percent of the total received by the university. Graduate Ph.D.s and postdocs such as Mary Betlach, Peter Lobban, and Janet Mertz left academia in droves to do research for private companies. Arthur Kornberg and Paul Berg, once resolute in their support of tradition, eventually formed private companies of their own. Stanley Cohen provides the most extreme example of the shift into private bioscience research. Despite later claiming deep reluctance, Cohen nevertheless vigilantly pursued virtually every private-venture opportunity that came his way: among other examples, he obtained a patent for his pharmaceutical database computer program, he was a co-applicant with Herbert Boyer on the first recombinant DNA patent, and he served as a consultant for a number of biotechnology companies, including Cetus. And there is the most famous example of Herbert Boyer using federal money to conduct the first cloning experiment at UCSF; he applied for and eventually received a patent on the technique, and then with venture capitalist Robert Swanson co-founded Genentech Corporation, the first genetic engineering company to use almost exclusively the technology that he had developed at a public university. The social and economic upheavals in the mid-1960s had ended and the consequences had reshaped the contours of a new kind of biosciences.[16]

As expected, not everyone agreed with the turn. Despite actions that would later speak to the contrary, Arthur Kornberg was one among many who nevertheless acknowledged at the time the "negative consequences of patent[ing] bioscience discoveries":

> It is a constraint of our capitalist system, that a patent precludes others from exploiting the discovery. . . . It results in restraints upon giving information that

you as a scientist are in good conscience required to provide. It's bad if you have something . . . and withhold it . . . because of a collaborative agreement with a company, or because it might affect the patentability of something that you are doing . . . , or because it might give an unfair commercial advantage to someone. . . . I'm not speaking of details of some . . . formula for Coca-Cola . . . I'm speaking of knowledge about a biological system that will have relevance to disease processes.

Still stung by what had happened at Asilomar, Congress, the state legislature, and Bay Area municipal governments prepared bills that would place more stringent controls on genetic engineering research. For instance, both the San Francisco Board of Supervisors and the Berkeley City Council expected to duplicate the actions of the Cambridge, Massachusetts, city council, which passed an ordinance that severely restricted recombinant DNA research. And the Bay Area's consummate populist Willie Brown led a massive—and wildly successful—campaign that prevented UCSF from building a new research laboratory on campus.[17]

But here again private industry provided a solution. The lax NIH guidelines that came out of Asilomar only applied to investigators who received federal money—private research lay outside the jurisdiction of federal regulation. Moreover, the ability of bioscientists who worked with recombinant DNA technologies to move their laboratories and research into more forgiving municipal districts furnishes a classical illustration of private industry's flexibility. For instance, to avoid any possibility that the Berkeley City Council might restrict recombinant DNA research, Cetus founders simply moved the company from Berkeley to Emeryville, a small industrial city only a few miles away from campus. Though Genentech maintained no laboratories in its first few years of operation, company founders nevertheless decided to elude all research regulations certain to come out of San Francisco city government and incorporate in nearby South San Francisco—known locally as the Industrial City. When Genentech scientists could not be found at company headquarters, they were most likely in the UCSF laboratories.

In the world of private bioscience research, there was much that was surprising. Who could have anticipated that investigators who had once flintily refused private money in the 1950s would embrace industry in the mid-1970s? That the country that had such deep abhorrence for runaway pure research would not also react against runaway private research? That the government that had become the largest sponsor of bioscience research the world has ever known would establish economic policies that protected and even encouraged private ownership of publicly supported discoveries? And who could deny that bioengineering—the application of recombinant DNA and cloning techniques toward the manipulation of genes—was the signature and lasting achievement of the biosciences in the postwar era, one likely to overshadow all the fundamental discoveries in its long-term historical consequences?[18]

This future may not have been clear in 1975, but it was evolving, mostly out of convenience. Indeed, the common alliances between bioscientists and private industry did not occur because both sides were desperate. Many of this new generation had become disillusioned with the academic establishment, the pretensions of prestigious university work, and the politics behind federal patronage that had so characterized the biosciences in the two decades after World War II. While their communitarian ideals would clash with efforts to maximize profits, private enterprise offered workplace autonomy, free-flowing research processes, and practical research agendas, all of which proved compatible with the humanitarian and libertarian ideals that underlay much of the patient-centered protests of the previous decade. And considering the amount of money they stood to make in private research, it should come as no surprise that a phrase like "doing good while doing well" would become a biotechnology company's unofficial anthem.[19]

Business was no unwitting accomplice either. Pharmaceutical and commercial biological firms had sponsored research for centuries, if not longer. But the unlimited potential of genetic engineering made this venture infinitely attractive: using a few recombinant DNA techniques, an investigator could conceivably combine any DNA segments to produce proteins that had enormous therapeutic, agricultural, or environmental utility. Venture capital, which had already gotten off to a roaring start in computers and electronics, added emolument to the industry's growth by providing—sometimes all too willingly—much needed start-up revenue. Moreover, to ensure that American corporations would lead in the development of genetic engineering—and not the Soviet Union—and to right the nation's sinking economic ship, Congress relaxed FDA regulatory laws to speed up product-to-market research and development time, liberalized patent laws to make them both stronger and more flexible, and slashed corporate and capital-gains taxes; municipal governments in the Bay Area extended tax credits and relaxed environmental restrictions; and the NIH sponsored Recombinant Advisory Committee (RACs) meetings to help investigators focus more attention on potential growth areas in genetic engineering. All these efforts and many more helped revitalize the nation's economy, in part because a new biotechnology industry took immediate shape: by the end of 1976, annual equity invested in all private genetic engineering firms averaged $70 million; total equity invested in biotechnology by pharmaceuticals alone surpassed $800 million; and university research laboratories received more than $250 million annually to conduct genetic engineering experiments from private interests. At the same time, the NIH alone was sponsoring almost 150 projects, each one at an approximate cost of $20,000. And this was just the beginning.[20]

No individual scientist, venture capitalist, government agency, laboratory, or university can claim credit for leading this scientific and industrial revolution. The making of a biotechnology industry emerged not because government officials guided it, university administrators willed it, the people needed it, or because a few investigators had a clear vision of future possibilities. The industry emerged because private interests found commercial applications for what had been discovered earlier—in universities with public money. Furthermore, private industry found it easy to ignore public worries, provide much needed research patronage, dismiss federal regulations, and unify a badly divided field. In short, the industry emerged because enough bioscientists found it acceptable to "do good while doing well." Investigators who entered the biotechnology industry defended their decision as "what was best for society," and they were not defensive or embarrassed about it. Indeed, they were proud and enthusiastic, as if doing applied science and making money deserved acclaim and admiration. Working in the biotechnology industry was egalitarian, humanitarian, even patriotic.

Some would use the example of biotechnology as an example of what was right with capitalism. Many would conclude that the only way to ensure a dynamic and expansive industry committed to improving health was to encourage even more private research. Investigators should not be regulated or guided by federal agencies, or else they would have no experimental liberty in which to pursue new paths. "The biotechnology industry," said a Genentech official, "was destined to become a great and 'moral' industry." Certainly the new generation of biotechnologists was less interested in adding incrementally more fundamental knowledge. This was not about repeating tradition, nor copying the path paved by investigators before them, nor uncovering new truths about life. Instead, this was about finding practical applications of fundamental knowledge, and earning a lot of money in the process.[21]

The public pays a high price for the commercialization of applied bioscience research—with its secrecy, materialism, commercialism, rootlessness, anti-intellectualism, and fanatical pursuit of profit, not to mention deeper bioethical considerations and the limitations that patent law imposes on further innovation. But there is no denying the wonder of the biotechnology industry and the real human benefits of applied bioscience research; the bioengineering of life-saving hormones such as insulin four years after Asilomar is just one scientific denouement.

We live with these consequences today, and we choose our future from among them, too.

Notes

Preface

1. I do not mean to imply that sociological studies of science are unimportant or lack significance, only that a study that places the organization and output of bioscience research in historical context can be equally productive, even informative for such studies. A substantial body of literature on the sociology of science has accumulated in recent years, and spans a wide spectrum of analytic frameworks. Some have borrowed from E. P. Thompson's celebrated essay on eighteenth-century bread-riots to argue that a particular "moral economy" informed the organization of basic and applied bioscience research in the San Francisco Bay Area. The complex set of relations between bioscientists at all three research universities reveals a set of unstated customs and traditions, or, "moral economies." Certainly, the language of basic and applied bioscience research emerged out of individual investigator's own efforts to describe their work. But it also emerged because it offered real advantages within a highly competitive research environment (Steven Shapin, "The House of Experiment in Seventeenth-Century England," *Isis* 79 (1988), 373–404). Others disregard questions of power in the experimental workplace and instead conduct ethnographies of laboratory life. Commonly referred to as "production-oriented" studies, these historians, philosophers, and sociologists of science have shown quite convincingly a powerful relationship between scientists' experimental behavior and epistemologies of scientific knowledge. For instance, in his study of *Drosophila* geneticists, Robert Kohler has shown how the organization of laboratory research promotes distinctive workplace cultures, and how this culture in turn shapes experimental outcomes (Robert Kohler, *Lords of the Fly* (Chicago, 1994]). At the same time, scholars such as Daniel Greenberg have shown how political decisions and public policy shapes the organization of scientific research (Daniel Greenberg, *The Politics of Pure Science* (New York, 1971). For more studies of the curious sociological underpinnings of science and the scientific community, see Bruno Latour and Steve Woolgar, *Laboratory Life: The Social Construction of Scientific Facts* (Beverly Hills, 1979); Andrew Pickering, ed., *Science as Practice and Culture* (Chicago, 1992); and Karen Knorr-Cetina, "Tinkering Toward Success: Prelude to a Theory of Scientific Practice," *Theory and Society* 8 (1979), 347–76.

2. Among many articles in *Science* on the subject of pure and applied research, see Michael Reagan, "Basic and Applied Research: A Meaningful Distinction?" *Science*, 17 Mar. 1967, 1383–84. For the article that initiated the debate, see R. E. Marshak, "Basic Research in the University and Industrial Laboratory," *Science*, 23 Dec. 1966, 1521–22.

3. Albert Einstein, quoted by Reagan, "Basic and Applied: A Meaningful Dis-

tinction?" 1383–84. For an example of a comparative study between bioscience disciplines, such as naturalist and molecular biology, see Joel Hagen, "Naturalist, Molecular Biologists, and the Challenges of Molecular Evolution," *Journal of the History of Biology* 32 (1999), 321–41.

Chapter 1. The Setting, 1946 . . .

1. Michael Malone, *The Big Score* (New York, 1985).
2. David Hounshell and John Kenly Smith, *Science and Corporate Strategy, Du Pont, 1902–1980* (Cambridge, 1995), 366; Louis Galambos and Jane Sewell, *Networks of Innovation* (Cambridge, 1995).
3. *Stanford Daily,* 7 Nov. 1958, 3; Rebecca Lowen, *Creating the Cold War University* (Berkeley, 1997); Vettel, "Research Life" (Ph.D. diss., 2003), ch. 1.
4. Clark Kerr, *The Gold and the Blue: A Personal Memoir of the University of California, 1949–1967,* vol. 1 (Berkeley, 2001), 29 and 56–59.
5. Warren Weaver's diary, 14 Nov. 1947, RG 1.2, 205D, box 7, file 49, RAC; Kerr, *The Gold and the Blue,* 83–84.
6. Richard C. Atkinson, "Robert Gordon Sproul," *California Journal* (Nov. 1999), 1; Kerr, *The Gold and the Blue,* 16–22.
7. Angela Creager, "Wendell Stanley's Dream of a Free-Standing Biochemistry Department at the University of California, Berkeley," *JHB* 29 (1996), 331–60.
8. Robert Kohler, *From Medical Chemistry to Biochemistry* (Cambridge, 1982), 158–65, 288, 334.
9. Creager, "Wendell Stanley's Dream of a Free-Standing Biochemistry Department."
10. Angela Creager, *The Life of a Virus* (Chicago, 2002), 47–50.
11. Sally Smith Hughes, *The Virus: A History of the Concept* (New York, 1977), 89–90; Stanley used this phrase repeatedly in his public correspondence, in BANC, Wendell Stanley Papers, MSS 78/18 c, file: "Correspondence."
12. Wendell M. Stanley, "Isolation of Crystalline Protein Possessing the Properties of Tobacco-Mosaic Virus," *Science* (81 (1935, 644–45; Barclay M. Newman, "Giant Molecules: The Machinery of Inheritance," *Scientific American* 158 (1938), 337; George Corner, *History of the Rockefeller Institute,* 320; Michel Morange, *A History of Molecular Biology* (Cambridge, 1998), 62–65. Kay, *A Molecular Vision of Life,* 111.
13. President Sproul to Stanley, 20 Jan. 1947; Office of the Dean of the Graduate Division to President Sproul, 30 July 1946: all in BANC, 420/Bio-chem, Berkeley: Office of the Dean of the Graduate Division, file: Correspondence.
14. Loomis diary, 17 Apr. 1950, box 7, file 49, RG 1.2, 205D, RAC; "All This and 2.0 Too," *Daily Californian,* 25 Oct. 1948, 12.
15. President Sproul to Stanley, 20 Jan. 1947; Office of the Dean of the Graduate Division to President Sproul, 30 July 1946: all in BANC, 420/Bio-chem, Berkeley: Office of the Dean of the Graduate Division, file: Correspondence.

Chapter 2. Patronage and Policy

Note to epigraphs: Warren Weaver, "Free Science," *Science and Imagination* (New York, 1967), 10–14, quoting Weaver, "Free Science," *New York Times,* 1945; Stephen Strickland, *The Story of the NIH Grants Program* (New York, 1989), 32.

1. For example, see Pnina Abir-Am, "The Discourse of Physical Power and Biological Knowledge in the 1930s: A Reappraisal of the Rockefeller Foundation's 'Policy' in Molecular Biology," *Soc. Stud. Sci.* 12 (1982), 341–82; Robert Kohler, *Partners in Science: Foundations and Natural Scientists, 1900–1945* (Chicago, 1991); John A. Fuerst, Ditta Bartels, Robert Olby, et al., "Responses and Replies to P. Abir-Am: Final Response of P. Abir-Am," *Soc. Stud. Sci.* 14 (1984), 225–63; Lily Kay, *The Molecular Vision of Life* (Oxford, 1993).

2. David Hollinger notices a similar trend, but from a slightly different vantage point: "cultural change [toward scientific autonomy] was a mediated . . . contingent, historically specific . . . transition from Protestant culture to pluralism," due in large part to the increase of Jewish professors in American universities following World War II. (David Hollinger, *Science, Jews and Secular Culture* [Princeton, 1996], 15, 21).

3. Frederick Rudolph, *The American College and University: A History* (New York, 1965, reprint), ch. 9; Robert Kohler, "Science, Foundations, and American Universities in the 1920s," *OSIRIS*, 2nd series (1987), 137.

4. According to Oliver Zunz, the Rockefeller Foundation served as a centerpiece for America's "institutional matrix." Zunz, *Why the American Century?* (Chicago, 1998); Albert F. Schenkel, *The Rich Man and the Kingdom: John D. Rockefeller, Jr., and the Protestant Establishment* (Minneapolis, 1995); and Peter Collier and David Horowitz, *The Rockefellers: An American Dynasty* (New York, 1977). For an excellent discussion of the Rockefeller Foundation and its support of the biological sciences at CalTech, see Kay, *The Molecular Vision of Life*; B. D. Karl and S. N. Katz, "The American Private Philanthropic Foundation and the Public Sphere, 1890–1930," *Minerva* 19, no. 2 (summer 1981), 238; and Kohler, "Science, Foundations and the American Universities," 3.

5. Peter Kuznick, *Beyond the Laboratory* (Chicago, 1987), 46–48 and ch. 2; Daniel Kevles, *The Physicists* (Cambridge, 1995), 237–38.

6. Kay, *The Molecular Vision of Life*, 42–44 and ch. 3; Kohler, "Science, Foundations and the American Universities"; Schenkel, *The Rich Man.*

7. The Rockefeller Foundation named its new vision, "Science of Man," and Weaver was a frequent and outspoken advocate of this new direction (Kay, *A Molecular Vision of Life*, 45–50).

8. Michael Morange, *A History of Molecular Biology*, trans. M. Cobb, (Boston, 1998), 80–81.

9. Kay, *Molecular Vision of Life*, 49; Warren Weaver, "Molecular Biology, Origins of the Term," *Science* (May 1938), 591–92.

10. Nathan Reingold, ed., *The Sciences in the American Context: New Perspectives* (Washington, D.C., 1979); Kay, *Molecular Vision of Life*, 48–50.

11. Warren Weaver (WW) interview of Stanley, Rockefeller Institute, Princeton, 14 Nov. 1947, in RAC, 205 D, 1.2, 205, 7, 49, file: UC, Virus Research (equipment, Wendell M. Stanley), 1947–1950.

12. Evaluation of grant RF 48132, 11/30–12/1/48, 1947–50, in RAC, 205 D, 1.2, 205, 7, 49, file: UC, Virus Research (equipment, Wendell M. Stanley).

13. BVL grant evaluation RF 52114, 20 June 1952; grant RF 48132, 30 Nov. 1948–1 Dec. 1948, both found in RAC, 205D, 1.2, 205, 7, 49, file: UC, Virus Research (Wendell M. Stanley), 1947–1950.

14. Rockefeller Foundation (RF) interviews of Stanley, including 14 Nov. 1947, in RAC, 205D, 1.2, 205, 7, 49, file: UC, Virus Research (equipment, Wendell Stanley), 1947–1950.

15. RF interview of Stanley, 14 Nov. 1947, in RAC, 205D, 1.2, 205, 7, 49, file: UC, Virus Research (equipment, Wendell Stanley), 1947–1950.

16. Stanley to WW, 8 Jan. 1948, in RAC, 205D, 1.2, 205, 7, 49, file: UC, Virus Research (equipment, Wendell Stanley), 1947–1950.

17. Weaver diary entry, 9 Jan. 1948, in RAC, 205D, 1.2, 205, 7, 49, file: UC, Virus Research (equipment, Wendell Stanley), 1947–1950; RS Morison to WW, 30 Jan. 1948, with WW 9 Jan. 1948 memo attached, in RAC, 205D, 1.2, 205, 7, 49, file: UC, Virus Research.

18. WW to Stanley, 10 Feb. 1948, in RAC, 205D, 1.2, 205, 7, 49, file: UC, Virus Research (equipment, Wendell Stanley), 1947–1950.

19. WW to RBF, 8 Mar. 1948, in RAC, 205D, 1.2, 205, 7, 49, file: UC, Virus Research (equipment, Wendell Stanley), 1947–1950.

20. Stanley to WW, 17 Sept. 1948, in RAC, 205D, 1.2, 205, 7, 49, file: UC, Virus Research (equipment, Wendell Stanley), 1947–1950.

21. Weaver's secretary, unknown, 15 Oct. 1948; Stanley to WW, 2 Nov. 1948; Weaver to Stanley, 9 Nov. 1948; WW to President Sproul, 8 Dec. 1948; WW diary entry, 25 Oct. 1950: all in RAC, 205D, 1.2, 205, 7, 49, file: UC, Virus Research (equipment, Wendell Stanley), 1947–1950.

22. Horace Judson, *The Eighth Day of Creation* (New York, 1979); see also Kay, *Molecular Vision of Life.*

23. The Rockefeller Foundation's initial positive reviews of the BVL strengthens Lily Kay's argument that Rockefeller patronage helped launch molecular biology, while Pnina Abir-Am's critique also seems appropriate because up-and-coming research programs such as the BVL found that the foundation's money was difficult to obtain. Indeed, a significant portion of Rockefeller money during this era went to established programs, such as the laboratories run by Linus Pauling and George Beadle at CalTech, or the Pasteur Institute at the Centre National de la Recherche Scientifique in France. The limits of Rockefeller influence on molecular biology are discussed by Abir-Am, "The Discourse of Physical Power and Biological Knowledge in the 1930s"; see also Kay, *Molecular Vision of Life.*

24. There is an enormous amount of literature that shows an undeniable relationship between the cold war and the rise of American science. See, for instance, Stuart Leslie, *The Cold War and American Science* (New York, 1993); Lowen, *Creating the Cold War University*; and Peter Galison and Bruce Hevly, eds., *Big Science* (Stanford, 1992).

25. For example, see Jessica Wang, *American Science in an Age of Anxiety: Scientists, Anticommunism, and the Cold War* (Chapel Hill, 1999).

26. Du Vigneaud to FBH, 27 Oct. 1944, in series 205, RF 1.1, box 9, folder 125/6, file: Stanford University, Chemistry—Loring, Hubert S., 1939 and 1941–48; Kay, *Molecular Vision of Life*, 187, 205, and 223–24, respectively.

27. Led by Robert Wiebe, Samuel Hays, and Louis Galambos, a school of revisionist historians have studied the rise of federal-professional relations, and the growth of administrative, state-supported scientific research. On the Department of Agriculture in the nineteenth century, the Depression, and war, see Victoria Harden, *Inventing the NIH* (Baltimore, 1986), esp. the epilogue; and Stephen Strickland, *Politics, Science and Dread Disease* (Cambridge, 1972), ch. 2. Brian Balogh, "Reorganizing the Organizational Synthesis: Federal-Professional Relations in Modern America," *Studies in American Political Development* 5 (spring 1991). Lois Magner, *A History of Medicine* (New York, 1992), 353–54.

28. For an excellent discussion of the OSRD's overhead policy and the precedent that it established, see Lowen, *Creating the Cold War University*, 58–66, esp. ch. 2; Nancy Rockafellar, "Interviews with W. F. Ganong" (San Francisco, 1995);

John McFee, "Medical School to Emphasize Research and Development," *Stanford Daily*, 21 May 1954.

29. Wang, *American Science in an Age of Anxiety*, 26–27.

30. Wang, *American Science in an Age of Anxiety*, 27, 30–31; Kevles, *The Physicists*, 348; Strickland, *Politics, Science and Dread Disease*, 20–21.

31. Kevles, *The Physicists*, 347, 397; Vannevar Bush, *Science: The Endless Frontier* (Washington, D.C., 1945).

32. Wang, *American Science in an Age of Anxiety*, 253–62; Kevles, *The Physicists*, 343. For a rich examination of the NSF, see Dian Belanger, *Enabling American Innovation* (West Lafayette, 1998); and Tony Appel, *Shaping Biology* (Baltimore, 2000).

33. Appel, *Shaping Biology*, 34.

34. Wang, *American Science in an Age of Anxiety*; Thomas Hager, *Force of Nature* (New York, 1995). J. Merton England, *A Patron for Pure Science* (Washington, D.C., 1982).

35. Statistics on the growth of science policy can be found in a number of different sources: Lowen, *Creating the Cold War University*, 2 and 121; Belanger, *Enabling American Innovation*, 47–49; Matthew Crenson and Francis Rourke, "By Way of Conclusion: American Bureaucracies . . ." in Louis Galambos, ed., *The New American State: Bureaucracies and Policies since World War II* (Baltimore, 1987), 147; Dickson, *The New Politics of Science*, intro. and ch. 1; Don K. Price, *The Scientific Estate* (New York, 1968); Daniel Greenberg, *The Politics of Pure Science* (Chicago, 1999); Detlev Bronk, "Science Advice in the White House," *Technology in Society* 2, nos. 1–2 (1980), 245–56; Margaret Rossiter, "Science and Public Policy Since World War II," *Osiris* 1 (1985), 273–94; Strickland, *The Story of the NIH Grants Programs*, 2; Harden, *Inventing the NIH*; and Appel, *Shaping Biology*, 31–32. On 14 percent, see Walter McDougall, *The Heavens and the Earth* (Baltimore, 1985), 462.

36. Kevles, *The Physicists*, 341–71, and Kevles, "K_1S_2," in Galison and Hevly, eds., *Big Science*, 312, 320.

37. Harden, *Inventing the NIH*, esp. 179, 180, 183, and epilogue; Strickland, *Politics, Science and Dread Disease*, 27–29, 49, 123. *NIH Factbook; Guide to National Institutes of Health Programs and Activities* (Chicago, 1976).

38. All quotes from Strickland, *Politics, Science and Dread Disease*, 120–25.

39. Italics added for spoken context; Strickland, *Politics, Science and Dread Disease*, 120–25.

40. According to one reviewer, in 1948, "it wasn't anything to travel 200,000 miles a year" to complete on-site reviews. Donald Swain, "The Rise of a Research Empire: NIH, 1930 to 1950," *Science* 14 Dec. 1962, 1233–37; and James Cassedy, "Stimulation of Health Research," *Science* 28 Aug. 1964, 897–902. Strickland, *The Story of the NIH Grants*, 25, 30, 45; Harden, *Inventing the NIH*, 183 and epilogue.

41. Philip Abelson, "A Critical Appraisal of Government Research Policy," in *Robert A. Welch Foundation Research Bulletin* no. 14 (Houston, Nov. 1963), 5.

42. James Patterson, *Grand Expectations* (New York, 1996), 122. Kevles, *The Physicists*, 359. Belanger, *Enabling American Innovation*. Cassedy, "Stimulation of Health Research," 898. Harvey Sapolsky, "Academic Science and the Military: The Years Since the Second World War," in *The Sciences in the American Context: New Perspectives*, ed. Nathan Reingold (Washington, D.C., 1979).

43. BVL follow-up evaluation for grant RF 54001, 14 June 1956, in RAC, 200D, 1.2, 226, 2173, file: UC, Photosynthesis (Melvin Calvin); Warren Shields

to Stanley, 14 Apr. 1952, in Stanley Papers, BANC, Carton 14, File 21; Sally Smith Hughes, "John Lawrence, M.D." (Berkeley, 2000); Kerr, *The Gold and the Blue,* 93 and 149.

44. F. S. Smyth to the State Legislature, unknown date, in UCSF Archives, AR 90–56, S/M, Dean's Office Records, 1936–1989, carton 5 of 13, file: Organized Research—Cancer Research Institute, 1945–1948; see also Creager, "Wendell Stanley's Dream," 349.

45. On appropriations for UC, see Senate Bill 569, 1947, authored by Dilworth (Chapter 360, Stats 1947) by Governor Warren (22 May 1947). On the "emergency" declaration, see Chapter 360, CA Statutes 1947, SB 569; see also J. F. Rinehart to President Sproul, 2 Apr. 1947, in UCSF Archives, AR 90–56, S/M, Dean's Office Records, CRI, 1936–1989, carton 5 of 13, file: Organized Research—the CRI, 1945–1948. For Berkeley's special budget, see "Statutes of California, 1948; General Laws, Amendments to Codes, Resolutions and Constitutional Amendments," passed at the 1948 Regular Session of the Legislature, ch. 23, p. 34, line item 138.1. Creager, "Wendell Stanley's Dream of a Free-Standing Biochemistry Department," 349.

46. Lowen, *Creating the Cold War University,* 121; Kerr, *The Gold and the Blue;* Stanley to the American Foundation, 5 Nov. 1955, in Stanley Papers, BANC MSS 78/18 c, carton 14, file 5.

47. Creager, "Wendell Stanley's Dream of a Free-Standing Biochemistry Department," 351.

Chapter 3. The Promise and Peril of the BVL

Notes to epigraphs: Loomis Diary, 6 Feb. 1951, in RAC, RG 1.2, 205D, Box 7, file 49. Angela Creager, "Wendell Stanley's Dream of a Free-Standing Biochemistry Department at the University of California, Berkeley," *JHB,* 29, no. 3 (fall 1996), 344–45, f(n) 51.

1. "Time to Change," *Daily Californian,* 29 Oct. 1945, 8; also found without a citation at www.lib.berkeley.edu/BANC/Exhibits/Biotech/stanley.

2. "Minimalism," also discussed in Chapter 1, dominated the biological sciences in the twentieth century, an approach that replaced Darwinian naturalism as the primary experimental paradigm. To implement a minimalist approach, physicists blended nicely with chemists interested in physiochemistry, while biochemists studied "specificity" and "structural" problems of protein metabolisms and clinical nutrition. It is in this scientific context of "minimalism" that Stanley's experiments with tobacco-mosaic virus took off (Barclay Newman, "Giant Molecules: The Machinery of Inheritance," *Scientific American* 158 [1938], 337). Lily Kay, *Molecular Vision of Life* (New York, 1993), 111–12; Michel Morange, *A History of Molecular Biology* (Cambridge, 1998), 34–36, 64–66.

3. Kay, *Molecular Vision of Life,* Interlude I; Horace Judson, *Eighth Day of Creation* (New York, 1979), 70–79; Thomas Hager, *Force of Nature* (New York, 1995); Morange, *A History of Molecular Biology,* 30–39.

4. Morange, *A History of Molecular Biology,* 47 and 56–57; Judson, *Eighth Day of Creation,* 33–39, 130–31.

5. Office of the Dean of the Graduate Division to President Sproul, 30 July 1946, in BANC, 420/Bio-chem, Berkeley: Office of the Dean of the Graduate Division, Correspondence. Judson, *The Eighth Day of Creation,* 30. Some scholars object to Judson's claim that *JEM* was an obscure journal and use the total subscription rate of *JEM* as evidence of its popularity; however, they ignore the char-

acterization of its audience. Indeed, readers of *JEM* typically came from a preclinical or biomedical background, while those who conducted harder fundamental research questions typically did not subscribe to *JEM*, which supports Judson's original claim. (Morange, *A History of Molecular Biology*, 56–57).

6. The presence of a phage researcher was highly unusual for virology because phage organisms are less than one-ten-thousandth of a millimeter in length and required highly specialized research tools and techniques. However, during the 1930s, Stent studied under Delbruck—considered by many the leading phage researcher in the world—which made him one of the first biochemists trained to recognize the chemical make-up of phage viruses (Mark Adams, *Bacteriophages* [New York, 1959], preface and introduction). Creager, "Wendell Stanley's Dream of a Free-Standing Biochemistry Department"; Stanley Papers, in BANC, file: General Correspondence, Franklin—Stanley, box 8, file 61. Morange, *A History of Molecular Biology*, 41–44.

7. Creager, "Wendell Stanley's Dream of a Free-Standing Biochemistry Department"; Grant application RF 52114, review on 20 June 1952, in RAC 205 D, 1.2, 205, 7, 49, file: UC, Virus Research (equipment, Wendell M. Stanley), 1947–1950; *Oakland Post Enquirer*, 10 July 1948, in UCB, CU-5: 1948, 420-Biochemistry, also referenced in Creager, "Wendell Stanley's Dream of a Free-Standing Biochemistry Department"; Warren Weaver diary, 10 Nov. 1952, in RAC, RG 1.2, 205D, box 7, file 50.

8. Creager, "Wendell Stanley's Dream of a Free-Standing Biochemistry Department" 342; Nancy Rockafellar, "Conversations with Dr. Leslie 'Latty' Bennett, The Research Tradition" (San Francisco, 1992), 65; Robert Kohler, *From Medical Chemistry to Biochemistry* (Cambridge, 1982), 158–65. Kohler argues that biochemistry could adapt to virtually all subdisciplines, which actually weakened biochemistry as a whole, making its influence less concentrated in an age of specialization. According to Kohler, the more biochemistry shifted into the orbit of clinical and medical research, the more isolated the field became. Kohler's analysis is confirmed by the experience of biochemistry at Berkeley.

9. University of California, In Memoriam, 1988: 308 2n 1:12, 88, in C. H. Li Papers, UCSF Archives, MSS 88–9, Series I: Correspondence Sub-Series 2, Subject, carton 17 of 56, file: Lasker Foundation, 1961–71.

10. The anterior pituitary hormones purified and isolated by Choh Hao Li are Interstitial Cell Stimulator (1939) Luteotrophic (1942), Andrenocortic (ACTH; 1943) and Somatostatin Growth (1944). According to Dr. Condliffe, who is ironically the only other scientist to isolate and purify a hormone other than Li, the BVL staff and UC administrators ostracized Dr. Li because of his "race" (Asian) and birthplace (Communist China); there is no evidence in the archives that substantiates or weakens this claim. C. H. Li, ed., "Biography" (University of California Press), v, in C. H. Li Papers, 1913–1987, UCSF, box HRL 1:1, Hormone Research Laboratory, 1950–1980; Nancy Rockafellar, "Interviews with Lloyd 'Holly' Smith; Building a Research-Oriented Department of Medicine" (San Francisco, 1998), 17; oral interview and electronic correspondence with Dr. Peter Condliffe, 16 Feb. 2002 at the UCSF Archives; Kerr to President Sproul, 13 Mar. 1958, in UCSF, Department of the History of Health Sciences, AR 87–46, file: Institute of Experimental Biology; "Making the Hormone for Human Growth; Target for Biochemist Choh Hao Li," *Medical World News*, 21 Feb. 1969, 32; and Mrs. Drew, "The Health Syndicate: Washington's Noble Conspirators," *Atlantic* (December), referenced in *Atlantic* (February 1968), all in C. H. Li Papers, UCSF, MSS 88–9, Series I: Correspondence Sub-

Series 2, Subject, carton 17 of 56, file: Lasker Foundation, 1961–1971; Sally Smith Hughes, "Interviews with Horace A. Barker" (Berkeley, 1994), 19.

11. Hughes, "Interviews with Horace A. Barker," 19–20; Barker to President Sproul, 24 Oct. 1952, in Barker Papers, BANC, CU-467, box 6, file 34; Creager, "Wendell Stanley's Dream of a Free-Standing Biochemistry Department," 344.

12. Creager, "Wendell Stanley's Dream of a Free-Standing Biochemistry Department."

13. Hughes, "Interviews with Horace A. Barker," 19–20; Barker to Wellman, 27 Oct. 1952, and Stumpf and Barker, "Report to the Pitzer Committee," all in Barker Papers, BANC, CU-467, box 6, file 34.

14. Stumpf to Barker, "Night Before Election," and Stumpf and Barker, "Report to the Pitzer Committee," all in Barker Papers, BANC, CU-467, box 6, file: 34; Rockafellar, "Conversations with Dr. Leslie 'Latty' Bennett," 27.

15. Rockafellar, "Conversations with Dr. Leslie 'Latty' Bennett," 27 and 32; Barker to Sproul on 24 Oct. 1952, in Barker Papers, BANC, CU-467, box 6, file 34.

16. Roger to "Nook" (Barker), 16 Nov. 1952, in Barker Papers, BANC, CU-467, box 6, file 34.

17. Rockafellar, "Conversations with Dr. Leslie 'Latty' Bennett," 32; on reassignments and transfers, compare different years of the medical, preclinical, and fellowship staff listings in the University of California, Announcements of the Medical School (San Francisco: respective years); Barker to President Sproul, 29 Sept. 1952, in Barker Papers, BANC, CU-467, box 6, file 34; see also www.lib.berkeley.edu/BANC/Exhibits/Biotech/stanley.html.

18. Though the BVL never formally organized into "camps" or "teams," Peter Condliffe, who worked at the BVL, describes these informal working teams accordingly (Personal and electronic correspondence with Dr. Peter Condliffe, 16 Feb. 2002, conducted at the UCSF Archives); for examples of research at the BVL, see UC, Virus Research (equipment, Wendell Stanley), 1947–1950, in RAC, 205D, 1.2, 205, 7, 49; Creager, "Wendell Stanley's Dream of a Free-Standing Biochemistry Department." Creager, *The Life of a Virus*, ch. 3.

19. At the risk of historical and scientific hindsight, there is evidence that experimentalists at the BVL had ample opportunity to see the biological significance of DNA in their own protein studies. See, for instance, C. A. Knight, "The Nature of Some of the Chemical Differences among Strains of Tobacco Mosaic Virus," *JBC* 171 (1): 297–308. Heinz Fraenkel-Conrat, "Protein Chemists Encounter Viruses," *The Origins of Modern Biochemistry; A Retrospect on Proteins*, eds. P. R. Srinivasan, Joseph S. Fruton, John T. Edsall, vol. 325 (New York, 1979), 310–12. J. I. Harris and C. A. Knight, "Action of Carboxypeptidase on TMV," *Nature*, 170: 613. Franklin and Caspar, *Nature*, 177: 928; Heinz Fraenkel-Conrat, "The Reaction of Tobacco Mosaic Virus with Iodine," *Journal of Biological Chemistry*, 1955, 217: 373–81. R. C. Williams and H. Fraenkel-Conrat, "Reconstitution of Active Tobacco Mosaic Virus from its Inactive Protein and Nucleic Acid Components," *Proc. Nat. Academy of Science*, 41: 690–98; see also Conrat, "Degradation of Tobacco Mosaic Virus with Acetic Acid," *Virology*, 1957, 4:1–4.

20. For UC enrollments, see Recorder of the Faculties, *University of California—Statistical Addenda* (UC Press, corresponding years). See alphabetical list of donors listed as file titles in "Correspondence Files," in Stanley Papers, BANC, file: General Correspondence, by name.

21. There are many excellent books about the discovery of DNA's double-helix structure, its internal copying mechanism, and the challenge of decipher-

ing the genetic code: Horace Judson, *The Eighth Day of Creation.* James Watson has written a lively and polemic account of this event in *The Double Helix* (London, 1981). Francis Crick attempted to correct some of Watson's exaggerations and inaccuracies in his own book, *What Mad Pursuit* (New York, 1988). James D. Watson and Francis H. C. Crick, "A Structure for Deoxyribose Nucleic Acid," *Nature* 171 (1953), 738–40; James D. Watson and Francis H. C. Crick, "Genetical Implications of the Structure of Deoxyribonucleic Acid," *Nature* 171 (1953), 964–67.

22. Stanley to Gamow, 1 Apr. 1955, in Stanley Papers, BANC, file: General Correspondence—Gamow, box 8, file 77; Judson, *Eighth Day of Creation,* 152–53 and 300; Michelle Slade, "Building a Legacy; A Chronology of Northern California Achievement," *Northern California's Bioscience Legacy* (Oakland, 19), 1; Morange, *A History of Molecular Biology,* 65; Fraenkel-Conrat, "Protein Chemists Encounter Viruses," 309–20.

23. Fraenkel-Conrat, "Protein Chemists Encounter Viruses," 309–20.

24. Lederle correspondence, 3 Feb. 1953, and Merck correspondence, 13 Apr. 1956, in Stanley Papers, BANC, General Correspondence, carton 14, file 4, by company name; for comparison to earlier donations, see 24 July 1950 and 20 June 1952, carton 15, file 28.

25. The original members of Gamow's RNA Tie Club are alanine: George Gamow (Berkeley); arginine: Paul Doty (Harvard), Richard Ledley (Washington area), Martinas Ycas (Army Quartermaster Corps), Robley Williams (virologist at Berkeley), Alex Dounce (Rochester), Richard Feynman (CalTech), Melvin Calvin (Berkeley), Norman Simmons (UCLA), Edward Teller (Berkeley), Erwin Chargaff (Columbia), Nick Metropolis (Los Alamos), Gunther Stent (Berkeley); proline: James Watson (Cambridge), Leslie Orgel (CalTech), Max Delbruck (CalTech), Francis Crick (Cambridge), Sydney Brenner (Cambridge), and one open seat for alternating honorary member. Stanley was never invited to fill the club's honorary seat. Alexander Rich, "Gamow and the Genetic Code," George Gamow Symposium, held at George Washington University, 12 Apr. 1996; see also Harper, Parke, Anderson, eds., *Astronomical Society of the Pacific Conference Series,* vols. 129, 116, 121.

26. Barker to President Sproul, 29 Sept. 1952 and 24 Oct. 1952, and Barker to Dr. Wellman, 27 Oct. 1952, in Barker Papers, BANC, CU-467, box 6, file 34.

27. For an unmarked, undated mimeographed copy of the lyrics, see the Barker papers in BANC, CU-467, box 6, file 34.

28. Barker to Kornberg, 4 Jan. 1954 and 28 Jan. 1955, in Kornberg Papers, SUA, SC 359, Kornberg, 89–063, box 24, Correspondence, 1955.

29. Morange, *A History of Molecular Biology,* 45.

30. Perhaps a comparative study of leading interdisciplinary research facilities might reveal similar institutional patterns and rigid commitments to certain paradigms, and of course, delayed responses to emerging molecular views of life. Three scholars make occasional reference to the tendency to overemphasize protein research: Morange, *A History of Molecular Biology*; Judson, *Eighth Day of Creation*; and Kay, *The Molecular Vision of Life.*

31. Creager, "Wendell Stanley's Dream of a Free-Standing Biochemistry Department."

32. Oral interview and electronic correspondence with Dr. Peter Condliffe, 16 Feb. 2002, at the UCSF Archives. With the luxury of hindsight, the results challenge Stanley's original autocatalytic theory of protein replication. For more examples of research at the BVL, see file: UC, Virus Research (equipment, Wendell Stanley), 1947–1950, in RAC, 205D, 1.2, 205, 7, 49.

Chapter 4. The Ascent of Pure Research

Notes to epigraphs: Steve Baffrey, "Med School's Curriculum Attracts Critics Already," *Stanford Daily*, 13 Oct. 1959, 1; see also Bob Montgomery, "New Medical Center to Have 'University-Oriented' Focus," *Stanford Daily*, 18 Feb. 1959, 4; Lederberg to Kornberg, 23 Feb. 1959, in SUA, SC 359, Kornberg, 89–063, box 25, correspondence, 1959.

1. David Hollinger also notes the irony of the growing autonomy of science and its parallel dependence on public money. According to Hollinger, postwar-era science discarded the romantic image of isolated genius and argued to the public that they deserved federal money for research. At the same time, science also argued in favor of professional autonomy because their expert knowledge guaranteed that they could—indeed should—police themselves. Hollinger calls this "scientific pluralism." However, as this and previous chapters have shown, sectarian scientific culture became *more* entrenched in the early postwar period. At the broadest level, the scientific community merely traded romantic images of isolated genius for parochial subdiscipline mastery over a specialized slice of scientific content. Perhaps sectarian science serves two purposes: it protects scientific universalism and professional autonomy. In other words, expert knowledge within each subdiscipline is more comprehensive and detailed and therefore is less approachable (David Hollinger, *Science, Jews and Secular Culture* [Princeton, 1996]).

2. Annalee Saxenian, *Regional Advantage* (Cambridge, 1996); Rebecca Lowen, *Creating a Cold War University* (Berkeley, 1997); Stuart Leslie, *Cold War and American Science* (New York, 1993).

3. Saxenian, *Regional Advantage*, 22; Lowen, *Creating a Cold War University*, 30, 90, 166–69; Terman Papers, SUA, SC 0160, Biographical sketch, introduction; WW diary (2 Feb. 1951), in RAC, box 6, folder 40, 1.2, 205D.

4. The four full professors practicing biochemistry in the chemistry department in 1946 were Loring, Luck, Griffin, and Tatum. *First Hundred Years, Stanford University School of Medicine* (Stanford, 1958) (ref. 11), 19; "Review, Stanford chemistry-Loring, H., 1939, 1941–8," in RAC, RF 1.1, series 205, box 9, folder 125/6; Loring to Hanson (13 Sept. 1948 and 19 June 1945), "Stanford chemistry-Loring, H., 1939, 1941–8," in RAC, RF 1.1, series 205, box 9, folder 125.

5. David Campaigne, "An academic catalyst: The life and work of George Beadle" (Ph.D. diss., Indiana University, 1997); Lily Kay, *The Molecular Vision of Life* (Oxford, 1993), 125, 130, 199–213.

6. Susan Spath, "C. B. van Niel and the Culture of Microbiology, 1920–65" (Ph.D. diss., UC Berkeley, 1999); GRP diary, 18 Jan. 1948, and RF administrators to van Niel, 21 Apr. 1948, and Starr to Miller, 9 Aug. 1954, all in RAC, RF 1.2, series 205D, box 6, folder 42/3, "Stanford University, microbiology (van Niel, CB), 1947–53."

7. Peter Gallison and Bruce Hevly, eds., *Big Science* (Stanford, 1992); Saxenian, *Regional Advantage*; Lowen, *Creating the Cold War University*; Terman Papers, SUA, SC 0160, Biographical sketch, introduction; Kay, *Molecular Vision of Life*, 205.

8. "Special Committee on the Medical School," 28 May 1953, in Lederberg Papers, SC 186, box 3, folder 6, "Special Committee, 1953."

9. "Special Committee on the Medical School," 28 May 1953, and appendix 3, in SUA, Lederberg Papers, SC 186, box 3, folder 6, "Special Committee, 1953."

10. *Stanford Daily,* 16 and 17 Apr. 1953, "Letters to the Editor."

11. Stanford Medical Conference, 30 Jan.–1 Feb. 1956, and "Majority Report of Subcommittee on Premedical Curriculum," 21 Feb. 1955, in Lederberg Papers, SUA, SC 186, box 3, folder 3 "Stimulus Correspondence with the Commonwealth Fund"; "SUMS, Committee on Future Plans," Part I; "Why Should a University Have a School of Medicine?" as a portion of the report, "Stanford Medical Family," 1 Dec. 1953, in SC 186, Lederberg Papers, box 3, folder 4; *Stanford Daily,* 14 May 1954 and 9 Aug. 1956, 1 and 2 Apr. 1959, 1; Cutting to the Stanford Medical Family, "Stanford Medical School, Income and Expenditures Report, 1953–1957," 1 Nov. 1955, in SC 186, Lederberg Papers, box 3, folder 5, file: 1954–57, Budget, Memo 26; *Stanford Summer Weekly,* 9 Aug. 1956, 1–2, and 31 July 1957, "Huge Project Is Med Center"; Steve Baffrey, "Med School's Curriculum Attracts Critics Already," *Stanford Daily,* 13 Oct. 1959, 1.

12. *Stanford Summer Weekly,* 9 Aug. 1956; John McFee, "Move Based on Desire for More Academic Strength," *Stanford Daily,* 14 May 1956, 3, and 24 Jan. 1957, 1; Sally Smith Hughes, "Arthur Kornberg; Biochemistry at Stanford, Biotechnology at DNAX" (Berkeley, 1997), 18.

13. See, among others, Walter McDougall, *"The Heavens and the Earth": A Political History of the Space Age* (Baltimore, 1985).

14. Stephen Strickland, *Politics, Science and Dread Disease* (Cambridge, 1972), 108.

15. "Medical School Quits S.F.," 27 Apr. 1960, 1; John McFee, "Move Based on Desire for More Academic Strength," 14 May 1956, 3, see also 17 and 24 Jan. 1957 and 26 Sept. 1957, all in *Stanford Daily*; *San Francisco Chronicle,* 10 Sept. 1958.

16. "Dean Cutting Quits Med School Job," 18 Feb. 1958, 26 Sept. 1958, "Medical School Quits S.F.," 27 Apr. 1960, 1, all in *Stanford Daily.*

17. "Robert H. Alway Ending Career at Stanford Medical School," *Medical Center Memo,* October 1977. Bob Montgomery, "New Medical Center to Have 'University-Oriented' Focus," *Stanford Daily,* 18 Feb. 1959, 4. Alway, "The Stanford Medical School; the Second-Half Century," 24 Apr. 1959, text of speech in Alway Papers, Lane, SID-5, box 1, file: 1.1, "1959." Steve Baffrey, "Med School's Curriculum Attracts Critics Already," *Stanford Daily,* 13 Oct. 1959, 1. See also Dean Alway to Medical Faculty Members, 7 July 1958, "Existence," *Medical Center Memo,* in Dean Alway Collections, Lane, box 1, file: 1.1, "1959."

18. Jerry Rankin, "Medical System Future Uncertain," *Stanford Daily,* 29 Oct. 1959, 1. See also "Eisenhower Proposes Health Insurance Plan," *Stanford Daily,* 19 Jan. 1954, 1.

19. "Grants-In-Aid to Stanford Scientists," *Stanford Daily,* 24 May 1956, 5.

20. "Biophysics Lab To Be Built on Science Quad," *Stanford Daily,* 27 Nov. 1957, 1. See also *Stanford Daily,* 16 Jan., 26 Feb., 20 and 24 Sep., and 27 Nov. 1957. Perhaps bioscientists merely embraced what Rebecca Lowen describes as "academic entrepreneurialism," seen most conspicuously by physical scientists and engineers during the 1940s (Lowen, *Creating the Cold War University*).

21. *Stanford Daily,* 7 Jan. 1958, 15 Oct. 1958, 2 Apr. 1959, 6 Apr. 1960, and 22 Sept. 1961. On the $500 given for bedside skills, see *Stanford Daily,* 15 Jan. 1960 and 6 Apr. 1960.

22. Lyman Stowe, "The Stanford Plan of Education for Medicine," paper read at annual meeting of Association of American Medical Colleges, Philadelphia, Penn., on 14 Oct. 1958, text in Lane, Alway Papers, SIP-5, Dean Alway Collection, box 1; also reprinted in *The Journal of Medical Education,* 36, no. 6, June 1961. *Stanford Daily,* 8 Oct. 1959.

23. Terman Papers, SUA, SC-160, Series III, box 4, folder 6: "Biochemistry, 1956–1959 at Washington University, St. Louis."

24. Horace Judson, *The Eighth Day of Creation* (New York, 1979), 322; Hughes, "Arthur Kornberg"; Arthur Kornberg, *For the Love of Enzymes* (Cambridge, 1989); Michel Morange, *A History of Molecular Biology* (Cambridge, 1998), 231–33. For more on Kornberg's work, see Arthur Kornberg, "Biological Synthesis of Deoxyribonucleic Acid," *Science*, 131 (1960), 1503.

25. Morange, *A History of Molecular Biology*, 233; Hughes, "Arthur Kornberg."

26. Arthur Kornberg, Paul Berg, Melvin Cohn, David Hogness, Armin Kaiser, Israel R. Lehman, and Robert Baldwin. Of the seven full-time professors in Stanford's inaugural biochemistry department, all but Robert Baldwin came from Washington University, St. Louis. Alway to Terman, 7 May 1957, SUA, SC 160, series III, box 4, folder 6, file: Biochemistry, 1956–59 at Washington Univ., St. Louis. Kornberg to Dr. P. L. Desai in Bombay, India, 2 Sept. 1958, in SUA, SC 359, Kornberg, 89–063, box 24, correspondence, 1958. Kornberg to Alway, 30 Dec. 1958, in SUA, SC 359, Kornberg, 89–063, box 25, correspondence, 1959. Kornberg, *For the Love of Enzymes*, esp. ch. 6, and 182.

27. Hughes, "Arthur Kornberg," 22–28.

28. Robert Kohler argues that because biochemistry remained in medical school, the field often merged with clinical medicine. According to Kohler, a biochemist often "accommodated his professional role and ideology to the needs and expectations" to the physician's. However, Kornberg refused to allow this to happen at St. Louis, much as Stanley refused at Berkeley. Indeed, Kornberg was ready to stretch his wings when Terman came calling; he was head of a department which did not interest him and he was doing more teaching than he preferred. (Robert Kohler, *From Medical Chemistry to Biochemistry* [Cambridge, 1982]). Austin Scott, "Award Given for Basic Research in the Field of Hereditary Traits," *Stanford Daily*, 16 Oct. 1959, 1. Sally Smith Hughes, "Paul Berg, A Stanford Professor's Career in Biochemistry, Science Politics and the Biotechnology Industry" (Berkeley, 1997), 45. On cooperative corporate strategies, see Kay, *The Molecular Vision of Life*, 9–10.

29. Kornberg to Alway on 9 Apr. 1958, in SUA, SC 359, Kornberg, 89–063, box 24, correspondence, 1958. Kornberg to Alway, 15 Aug. 1958, in SUA, SC 359, Kornberg, 89–063, box 31, correspondence, file: "Stanford—July—Dec. '58" (Public Health Service grant applications transferred to Stanford: E-2433 for $64,625; RG, A, C, E, H—5918 for $139,710, and Biochemistry Training Grant for $48,600). Kornberg, *For the Love of Enzymes*, 182–83. WFL Diary, entry 2 Feb. 1951 in Series 205, RF 1.1, box 9, folder 127, file: Stanford University, Chemistry Loring, Hubert S. 1949–55; Kornberg to Alway, 19 Feb. 1959, in SUA, SC 359, Kornberg, 89–063, box 25, correspondence, 1959.

30. For a study of struggle within genetics, see Jan Sapp, "The Struggle for Authority in the Field of Heredity, 1900–1932: New Perspectives on the Rise of Genetics," *JHB*, vol. 16, 1983, 311–42. See also Morange, *A History of Molecular Biology*, 18 and 20–25.

31. Judson, *Eighth Day of Creation*, 271. On Lederberg's research in molecular genetics, see Joshua Lederberg, "A Fortieth Anniversary Reminiscence," *Nature*, vol. 324, 1986, 627–28; see also Morange, *A History of Molecular Biology*, esp. ch. 5.

32. Lederberg to Kornberg, 27 Jan. 1958, in SUA, SC 359, Kornberg, 89–063, box 31, correspondence, 1957.

33. Kornberg to Terman and Alway, 1 Apr. 1958, in SUA, SC 359, Kornberg,

89–063, box 24, correspondence, 1958. Lederberg to Kornberg, 27 Jan. 1958 and 2 June 1958, both in SUA, SC 359, Kornberg, 89–063, box 31, correspondence, 1957. Sally Hughes, "Arthur Kornberg; Biochemistry at Stanford," 20.

34. Dean Alway to Lederberg, 1 July 1958, in SUA, SC 359, Kornberg, 89–063, box 31, Correspondence, 1957.

35. Dean Alway to Wallace Sterling, Minutes of the Meeting of the Executive Committee of the Medical Council of Stanford University School of Medicine, 21 Jan. 1959, in Lane, SID-5, Dean Alway Collection, box 1, file: 1.1, 1959. *Stanford Daily*, 7 Jan. 1958, 15 Oct. 1958, 2 Apr. 1959, 6 Apr. 1960, 22 Sept. 1961.

36. Dean Alway to Sterling, 11 Jan. 1960, in Lane, SID-5, Dean Alway Collection, box 1.2, 1960.

37. Daniel Kurzman, *Disaster! The Great San Francisco Earthquake and Fire, 1906* (New York, 2001); *The Call*, *San Francisco Chronicle*, and *San Francisco Examiner*, all dated 19 Apr. 1906.

38. Julius Comroe and Richard Havel, eds., *Cardiovascular Research Institute: The First Twenty-five Years* (San Francisco, 1983), 6. Nancy Rockafellar, "Interviews with Clark Kerr" (San Francisco, 1997), 13.

39. Clark Kerr, *The Gold and the Blue* (Berkeley, 2001), ch. 3. Comroe and Havel, *Cardiovascular Research Institute.*

40. John Walsh, "Graduate Education: ACE Study Rates Departments Qualitatively," *Science* 152 (May 27, 1966): 1226–28. Kerr, *The Gold and the Blue*, 56–60. Rockafellar, "Interviews with Clark Kerr," 33–34; Nancy Rockafellar, "Interviews with Lloyd 'Holly' Smith" (San Francisco, 1998), 37 f(n) 30—documents held in UCSF, AR—92–83.

41. Kerr, *The Gold and the Blue*, 320; Rockafellar, "Interviews with Clark Kerr," 31.

42. Seymour S. Kety and Robert Forster, *Julius H. Comroe, Jr., 1911–1984: A Biographical Memoir* (Washington, D.C., 2001); Comroe and Havel, *Cardiovascular Research Institute*, 7. Nancy Rockafellar, "Interview with W. F. Ganong, MD in Neuroendocrinology in the Academic Medical School" (San Francisco, 1995), 20–22. Rockafellar, "Interviews with Lloyd 'Holly' Smith," 36. Rockafellar, "Interviews with Clark Kerr," 26.

43. Comroe and Havel, *Cardiovascular Research Institute*, 8–12; Kety and Forster, *Julius H. Comroe, Jr., 1911–1984; A Biographical Memoir.*

44. "The Ivory Tower; Here's How It All Began," *Synapse*, 27 Sept. 1957, 1. Comroe and Havel, *Cardiovascular Research Institute*, 10 and 22–23.

45. Kerr, *The Gold and the Blue*, 214–25; Rockafellar, "Interviews with Clark Kerr," 33–34. Joan Trauner, "Interview with John Saunders unpublished oral history," in UCSF, Saunders Papers, unsorted files, MSS 90–73, file: Interview transcripts, Apr. 1978, 10.

46. Kerr, *The Gold and the Blue*, 321. Trauner, "Interview with John Saunders unpublished oral history," in UCSF, Saunders Papers, unsorted files, MSS 90–73, file: Interview transcripts, Apr. 1978, 10.

47. Notes from 23 Sept. 1949 meeting, Dean Smyth's office with Saunders, Rinehart, and Anderson; Saunders's memo to President Sproul, 23 Jan. 1950, and Advisory Committee on Inter-Campus Medical Teaching to Sproul, 2 Oct. 1950, all in UCSF, AR 87–46, file: Institute of Experimental Biology. Trauner, "Interview with John Saunders unpublished oral history," in UCSF, Saunders' Papers, unsorted files, MSS 90–73, file: Interview transcripts, Apr. 1978, 6 and 10; William R. Lyons to J. D. C. M. Saunders, 24 Feb. 1950, in UCSF, Saunders' Papers, unsorted files, MSS 90–73. Saunders to President Sproul, 10 Sept. 1957, in UCSF, AR 87–46, file: Institute of Experimental Biology.

48. In 1957, the budget of the UC Anatomy Department was $117,561, of which $117,011 went to research at Berkeley while UCSF had $510 to cover instructional and administrative costs. There is also a rich source of material on Berkeley's unwillingness to accommodate the Metabolic Research Unit and any program peripherally related to it. For instance, Sproul backed Stanley's push of Herbert Evans, one of Berkeley's most productive and recognized bioscientists, toward retirement. Evans to Sproul, 25 June 1951 and 26 Oct. 1951, and Sproul to Saunders, 17 June 1953, and Saunders to President Sproul, memo, 10 Sept. 1957, all in Department of the History of Health Sciences, AR 87–46, file: Institute of Experimental Biology, see handwritten notes, "1950s Report," for private correspondence about the "BVL massacre." Rockafellar, "Interviews with Clark Kerr," 32.

49. Comroe to Saunders et al., 20 May 1960, in UCSF, AR 90–56, carton 6, file: Organized Research—Cardiovascular Research Institute 1958–60; see also UCSF, MSS 90–73, John Saunders' Papers (unsorted), file: 1963 Long Range Dept Goals. Trauner, unpublished and unsorted oral history of Saunders, 11, in file: Interview transcript 30, Oct., 1978, labeled as MSS 90–73. "The Cardiovascular Research Institute of the University of California School of Medicine, Statement of Purpose," 21 Feb. 1955, in AR 90–56, carton 6, file: Organized Research—Cardiovascular Research Institute 1950–57. Greenberg, "Distinguished Biochemical Discoveries and Biochemists on the Berkeley Campus, 1920–1970," unpublished report, in Perspectives in Biology and Medicine (UCSF publication), vol. 16, 1, autumn 1972. Rita Carroll and Nancy Rockafellar, "Conversations with Dr. Leslie 'Latty' Bennett; the Research Tradition at UCSF" (San Francisco, 1992), 78–79.

50. Comroe and Havel, *Cardiovascular Research Institute*, 21–22; Rockafellar, "Interviews with Clark Kerr," 25. See also UCSF, AR 86–32, Human Biology Program, 1955–84, carton 1, file 18.

51. Comroe to Saunders, 18 Nov. 1959, in UCSF, AR 90–56, carton 6, file: Organized Research—Cardiovascular Research Institute 1950–57.

52. Rockafellar, "Interviews with Rudi Schmid," 104. Rockafellar, "Interviews with Clark Kerr," 29, f(n) 10. *Report of the Committee on Impact of Affiliated Institutions*, 8 June 1961, in UCSF, AR 71–12, carton 1, UCSF Chancellors Office, 5A.

53. Among those currently at UCSF who witnessed the 1964 coup firsthand, a great majority believe without any hesitation that Saunders was *the* antagonist in early efforts to establish a basic bioscience research program.

54. The coup was set off when the petitioners sent a letter to Kerr on 20 Nov. 1964 asking for "an immediate future decision (regarding) certain high-level decisions of utmost importance to this campus." The petition was signed by Julius Comroe, B. C. Cullen, J. E. Dunphy, I. S. Idelman, R. M. Featherstone, A. Simon, L. H. Smith, M. Sokolow, W. O. Reinhardt, and A. Margulis (E. Page and E. Jawetz attended the meeting but did not sign the petition), to Clark Kerr, 27 Nov. 1964, is attached as an appendix to virtually every oral history conducted by the UCSF Oral History Program, including Rockafellar, "Interviews with Rudi Schmid," appendix D.

55. William Boquist, *San Francisco Examiner*, 21 Jan. 1965. See also *San Francisco Chronicle*, 20 Jan. 1965, and the *San Francisco News Call Bulletin*, 19 Jan. 1965. Rockafellar, "Interviews with Lloyd 'Holly' Smith," 29–32, and f(n) 19; see also oral history interviews by Rockafellar with Clark Kerr, Morton Meyer, Alex Margulis, Richard Havel, and others.

56. Kerr, *The Gold and the Blue*, 214–25; Rockafellar, "Interviews with Clark

Kerr." Rockafellar, "Interviews with Rudi Schmid," 104, f(n) 25 and appendix D. Regular Meeting before the Academic Senate, 2 June 1966, in UCSF, AR 90–56, carton 6, file: Organized Research—Cardiovascular Research Institute 1958–60.

57. Comroe and Havel, *Cardiovascular Research Institute,* 26.

58. Loomis diary, 17 Apr. 1950, in RAC, RG 1.2, 205D, box 7, file 49.

59. EWS [Strong] cc'ing Beck, Millman, Miller, and Owen, memo, 24 Jan. 1962; EWS to Beck et al., memo 25 Jan. 1962, both in BANC, CU-149, box 75, also listed as CU-149, 695, 75:33, Molecular Biology, Proposed.

60. Seaborg to Strong, notes from telephone conversation, 6 Feb. 1962, written by KCM, in BANC, CU-149, Box 75, also listed as CU-149, 695, 75:33, Molecular Biology, Proposed.

61. Strong to Kerr, 15 May 1962 and 28 May 1962, in BANC, CU-149, box 75, also listed as CU-149, 695, 75:33, Molecular Biology, Proposed. Strong to Williams, 5 Apr. 1962, in BANC, CU-149, Box 75, also listed as CU-149, 695, 75:33, Molecular Biology, Proposed. Strong cc'ing Beck, Millman, Miller, and Owen, memo 25 Jan. 1962, in BANC, CU-149, Box 75, also listed as CU-149, 695, 75:33, Molecular Biology, Proposed. Kerr to Strong, 14 Feb. 1962, in BANC, CU-149, Box 75, also listed as CU-149, 695, 75:33, Molecular Biology, Proposed. Strong to Mazia, 5 Apr. 1962, in BANC, CU-149, Box 75, also listed as CU-149, 695, 75:33, Molecular Biology, Proposed. Miller and Mauchlan to Seaborg, memo 19 May 1962, in BANC, CU-149, box 75, also listed as CU-149, 695, 75:33, Molecular Biology, Proposed.

62. Strong to Mazia, 5 Apr. 1962, in BANC, CU-149, box 75, also listed as CU-149, 695, 75:33, Molecular Biology, Proposed. Miller to Dean Connick, memo with Calvin, 6 Aug. 1962, in BANC, CU-149, box 75, also listed as CU-149, 695, 75:33, Molecular Biology, Proposed.

63. On Lerner, see 12 Nov. 1962; on Tinoco, see 12 Nov. 1962; on Calvin, see 13 Nov. 1962; on Machlis, see 19 Nov. 1962; and see especially letters to Chancellor Strong, 18 Dec. 1962 and 19 Dec. 1962, in BANC, CU-149, box 75, also listed as CU-149, 695, 75:33, Molecular Biology, Proposed.

64. Alden Miller, memo, 7 Feb. 1962, in BANC, CU-149, Box 75, also listed as CU-149, 695, 75:33, Molecular Biology, Proposed.

65. Seaborg to Strong, notes from phone call on 6 Feb. 1962; KCM memo with Alden Miller's handwritten note at the bottom, in BANC, CU-149, box 75, also listed as CU-149, 695, 75:33, Molecular Biology, Proposed.

66. McEntire to Strong, 5 June 1962, in BANC, CU-149, Box 75, also listed as CU-149, 695, 75:33, Molecular Biology, Proposed. Notes from Kerr/Strong meeting, 15 Jan. 1963, in BANC, CU-149, Box 75, also listed as CU-149, 695, 75:33, Molecular Biology, Proposed.

67. EWS memo, 30 Apr. 1963, and Reed memo 9 May 1963, both in BANC, CU-149, box 75, also listed as CU-149, 695, 75:33, Molecular Biology, Proposed. Miller to Kerr, 18 June 1962, in BANC, CU-149, box 75, also listed as CU-149, 695, 75:33, Molecular Biology, Proposed.

68. "Dr. Karl F. Meyer Honored," *Synapse,* 25 Apr. 1958, 1.

69. Susan Wright, "Recombinant DNA Technology and its Social Transformation, 1972–1982," *OSIRIS,* 2nd series, 1986, 2: 303–60. For a historiographic view, see Robert Hynes and Philip Hanawalt, *The Molecular Basis of Life* (San Francisco, 1968) and David Freifelder, *Recombinant DNA* (San Francisco, 1978). *Stanford Daily,* 16 Oct. 1959, 1.

70. Kay, *The Molecular Vision of Life,* 16.

Chapter 5. Research Life!

Note to epigraph: Daniel Greenberg, "Money for Research," *Science*, 19 May 1967, 922.

1. A few scholars have addressed the transformation of the biological sciences during the last half of the twentieth century. Most link changes in scientific focus with the errors of administrative leadership, but this historical approach tends to focus too much on the failures of elites. Certainly, in the case of Berkeley, UCSF, and Stanford, the rigid commitments to disciplinary divisions, and the choices made by administrators and individual investigators, dramatically weakened the foundations upon which they had built their impressive empires. But the cracks in the walls of the ivory tower did not mean change was inevitable, nor did it determine the direction that change might take—these organizational weaknesses merely made the empire susceptible to outside pressure. Indeed, the emerging protest culture capitalized on disciplinary divisions and bent them to their favor. For patient-centered analysis of the biological sciences and biomedicine, see Kenneth Ludmerer, *Time to Heal* (Oxford, 1999); David Rothman, *Strangers at the Bedside* (New York, 1991). For other surveys of American medicine in the twentieth century, see Paul Starr, *The Social Transformation of American Medicine* (New York, 1982); Rosemary Stevens, *In Sickness and in Wealth*; (reprint, Baltimore, 1999).

2. Dael Wolf, *Scientific American*, July 1965, vol. 213, no. 1; Daniel Kevles, *The Physicists* (Cambridge, 1995), 387. Walter McDougall, *The Heavens and the Earth* (Baltimore, 1985). Harold Orlans, *Science*, 10 Feb. 1967, 665. *The NIH Factbook: Guide to National Institutes of Health Programs and Activities* (Chicago, 1976), esp. 99.

3. Transcripts of KQED's broadcast in the Stanley papers, BANC MSS 78/18 c, carton 14, folder 8, and commentary in a letter to the *San Francisco Chronicle*, 25 June 1958, 4. *Stanford Daily*, 6 Feb. 1953, 25 Jan. 1957, 1 Oct. 1957, 29 Oct. 1958. Kevles, *The Physicists*, 387, who quotes Klaw Spencer, *The New Brahmins: Scientific Life in America* (New York, 1968), 35.

4. Julius Comroe, "Research and Medical Education," *Journal of Medical Education*, vol. 37, no. 12, Dec. 1962, in UCSF archives, MSS 94–54, Rutter papers, Box 5 of 13, Series II, File: School of Medicine—Executive Committee, 1971–1972. Wendell Stanley, "Presentation to the American Foundation," delivered in New York City on 5 Nov. 1955, in the Stanley Papers, BANC, MSS 78/18C, carton 14, file 5. Tuesday Lecture Series, *Stanford Daily*, 3 July 1953.

5. "FAS preamble," in the FAS Papers, Special Collections Research Center, University of Chicago Library, box XX, folder 4, file: Jan. 1946. Jessica Wang, *American Science in an Age of Anxiety* (Chapel Hill, 1999); Kevles, *The Physicists*, esp. ch. 22.

6. Many tend to believe that the nationalistic and ideological pressures of the cold war era handcuffed the political influence of the FAS. While such an assessment might be true for physicists and chemists whose professional autonomy eroded under the heavy hand of Draconian federal agencies such as HUAC, it badly misrepresents the scientific authority and autonomy of biological scientists in the Bay Area. Perhaps the most compelling study of science during the cold war era is Jessica Wang, *American Science in an Age of Anxiety*. See also Peter Kuznick, *Beyond the Laboratory* (Chicago, 1986); Elizabeth Hodes, "Precedents for Social Responsibility Among Scientists: The American Association of Scientific Workers and the Federation of American Scientists,"(Ph.D. diss., UCSB, 1982);

Barton Bernstein, "Origins of the U.S. Biological Warfare Program," in Susan Wright, ed., *Preventing a Biological Arms Race* (Cambridge, 1990), ch. 1.

7. Wang, *American Scientists in an Age of Anxiety*, 59. *Time*, 20 Aug. 1945. Michael Amrine to M. Kasha, 1 Oct. 1946, in FAS Papers, Special Collections Research Center, University of Chicago Library, box XX, folder 5.

8. Charles Wagner to Dr. Walter Murphy, and NCAS press release on 29 Sept. 1946, in FAS Papers, Special Collections Research Center, University of Chicago Library, box XX, folder 5. Raymond Lawrence, *Oakland Tribune*, 4 Mar. 1946, in FAS Papers, box XX, folder 4.

9. Kirk dropped out of the FAS soon after it was formed. Preamble signatories, 15 Mar. 1947, and *Congressional Quarterly*, 15 Mar. 1947, vol. V, no. 11, 89, all in FAS Papers, Special Collections Research Center, University of Chicago Library, box XX, folder 2 and 4. Addis (perhaps of the Stanford University School of Medicine) to Miss Shuler, 27 Nov. 1945, in FAS Papers, box XX, folder 4. Membership list in the NCAS (pamphlet), 1 Aug. 1946, in FAS Papers, Special Collections Research Center, University of Chicago Library box XX, folder 6.

10. On Berkeley enrollments, see 1972 Addendum, in FAS papers, box 40, folder: Berkeley, California. Jack Hollander to Daniel Singer of the D.C. FAS, 12 Jan. 1962; on Stanford membership list, see James Gill to the D.C. office, "Chapter Dues and Membership Records Form," 22 Dec. 1959, in FAS Papers, Special Collections Research Center, University of Chicago Library, box 41, folder: Stanford Chapter, 1950s. Ron Stein, "Socialized Medicine in England, the National Health System," *Synapse*, 6 Oct. 1958, 4. Elvyn Cowgill and Dr. Seymour Farber, "Symposium On the Family Explores Pressures," *Synapse*, 15 Jan. 1964, 1. Marian Knox to Lederberg, 22 Sept. 1975, in Lederberg Papers, SUA, SC 186, box 1, file: Affirmative Action, 1972–75. Schachman to President Sproul, 1 Nov. 1950, posted on the BANC Web site: http://sunsite.berkeley.edu/uchistory/archives_exhibits/loyaltyoath/schachman.jpg. (The Berkeley biotechnology Web site unfortunately does not label the archival location of the documents posted.) Peter Condliffe interview transcripts, from 1 Aug. 2001 (Condliffe is currently an emeritus member of the NIH and was a graduate student at Berkeley under C. H. Li and Gunther Stent). Though it is always a challenge to prove a negative, nothing in the Wendell Stanley Papers suggests he was politically "liberal," as Nicolas Rasmussen claims in *Picture Control* (Stanford, 1997), 199. Stanley and Pauling correspondence, 21 May 1957 and 6 Nov. 1957, in Stanley Papers BANC, box 11, file 59.

11. Judson, *The Eighth Day of Creation.* Wang, *American Science in an Age of Anxiety.* Philip Abelson's quotation adapted from Kevles, *The Physicists*, 394–95, f(n) 4, who quotes Ralph Lapp, *The New Priesthood* (New York, 1965), 30.

12. Alan Waterman of the NSF projected 35% growth rates, while Philip Handler and Frederick Seitz independently predicted that half of the GNP would be devoted to research by the year 2000. Kenneth Pitzer, "How Much Research?" *Science*, 18 Aug. 1967, 779. Hughes, "William J. Rutter, Ph.D.," 77; D. S. Greenberg, "LBJ's Budget," *Science*, 27 Jan. 1967, 434.

13. Heinz Fraenkel-Conrat, "The Role of the Nucleic Acid in the Reconstitution of Active Tobacco Mosaic Virus," *Journal of the American Chemistry Society*, vol. 78, 1956, 882–83. Lederberg, "Genes and Antibodies," *Science*, 1959, 1649–53 and *Stanford Medical Bulletin* 19, 1961, 53–61 postscript 3/61. Kornberg, "Never a Dull Enzyme," *Annual Review of Biochemistry*, vol. 58, 1989, 1–30; Kornberg, *For the Love of Enzymes* (Cambridge, 1989).

14. Nancy Rockafellar, "Interviews with W. F. Ganong" (San Francisco, 1995), 34–36.

15. Daniel Mazia to Connick, 2 Dec. 1966, BANC, CU-149, 695 T, 75:34, Molecular Biology. Adriano A. Buzzati-Traverso, "Scientific Research: The Case for International Support," *Science*, 11 June 1965, 1440. Julius Comroe, *Cardiovascular Research Institute: The First Twenty-Five Years; 1958–1983*.

16. Certainly not all of Lederberg's preoccupations were frivolous. For instance, Lederberg requested support from the NIH to create a computer network system that would centralize all of the medical school's files into a single database. The NIH gave him $555,531 to start the ACME real-time computing laboratory—a precursor to the development of the Internet. Joshua Lederberg Papers, SUA, ACME Project, SC 236, box 1, file: Stanford Medical Computing, Before ACME, and ACME Budget. Dean Alway to Sterling, 13 July 1962, in Dean Alway Collection, SUMC Archives, S1D5, box 1.4, file: 1962. Also, interview with Robert Glaser, Dean of Medicine at the SUMC.

17. Harold Orlans, "Developments in Federal Policy Toward University Research," *Science*, 10 Feb. 1967, 665. Kevles, *The Physicists*, referencing Fischer, "Why Our Scientists Are About to Be Dragged, Moaning, into Politics," *Harper's* (Sept. 1963), 16. David Greenberg, "Money for Research: LBJ's Advisers Urge Scientists to Seek Public Support," *Science*, 19 May 1967, 920.

18. *Stanford Daily*, 12 May 1953, 18 May 1953, 1 May 1960, 24 Feb. 1961. *Synapse*, 26 Oct. 1960, 10. Patterson, *Grand Expectations*, 408–9. Jim Lieberman, "An Open Letter to My Draft Board," *Synapse*, 23 May 1958, 3.

19. Arthur Daemmrich, "A Tale of Two Experts: Thalidomide and Political Engagement in the United States and West Germany," *The Society for the Social History of Medicine*, vol. 15, no. 1, 137–58. Phocomelia deformities include stunted arms and legs, misshapen hands and feet, hands or feet that appear to grow directly from the torso, and severely damaged internal organs.

20. Daemmrich, "A Tale of Two Experts," 137. Robert Cutter to Wendell Stanley, 1 Apr. 1958, and Robert Cutter to Dean Grether, 9 Nov. 1959, both in Stanley Papers, BANC, carton 14, folder 39, correspondence. The prosecuting attorney was a young Melvin Belli.

21. G. Burroughs Mider to Wendell Stanley, 23 Feb. 1960, in Stanley Papers, BANC, carton 16, file 18; James Patterson, *The Dread Disease: Cancer and Modern American Culture* (Cambridge, 1987), 201–30. *Washington Post*, 15 July 1962, 1, 8. Robert Crain, *The Politics of Communal Conflict* (Indianapolis, 1969). Joshua Blu Buhs, "The Fire Ant Wars," *Isis*, 2002, 93:377–400, esp. 391. Donna Hamilton, "1961: Spurring Drug Reforms to Prevent Birth Defects" (*The History of the FDA*, November 1997), on FDA Web site: www.fda.gov. Daemmrich, "A Tale of Two Experts," 137–58.

22. Rachel Carson, *Silent Spring* (196; reprint, Boston, 1994). Throughout 1963, Bay Area newspapers routinely listed Carson's book as a regional bestseller in nonfiction; Charles Reich, *The Greening of America* (New York, 1971); Lewis Mumford, *The Myth of the Machine* (New York, 1967); Thomas Kuhn, *The Structure of Scientific Revolutions* (19; reprint, Chicago, 1996); J. D. Bernal, *The Social Function of Science* (Cambridge, 1967). Background on Thomas Parkinson and H. Bruce Franklin's courses comes from conversations with David Kessler in the Bancroft Library at Berkeley and Margaret Kimball, the Stanford University archivist.

A number of scholars have made a compelling case that environmental concerns of those living in the Bay Area were rooted in the vast social changes that took place in that region after World War II. The historical timing of, among other changes, a booming population and a rising standard of living in the 1940s caused unusual stresses upon the region's natural resources and led to an expanding interest in outdoor recreation in the next decade. For a discussion of science and environmental affairs in California during the postwar period, see Samuel Hays, *Beauty, Health and Permanence* (Cambridge, 1987), 174–75.

23. Rhodri Jeffreys-Jones, *Peace Now!* (New Haven, 1999), and Donald Phillips, *Student Protest, 1960–1970* (Lanham, Md., 1985), 166.

24. Morales to Reinhardt, 21 Oct. 1963, in AR 90–56, UCSFA, carton 12, file: 24–2, Biochemistry Search Committee; Joshua Lederberg, *Science*, 20 Oct. 1967, 313. Hughes, "Arthur Kornberg, M.D.; Biochemistry at Stanford, Biotechnology at DNAX," 40. Heinz Fraenkel-Conrat to the Staff of Molecular Biology, 27 Oct. 1966, in BANC, file: 34:15 Molecular Biology: Correspondence, 1965–75. Julius Comroe, "Research and Medical Education," *Journal of Medical Education*, vol. 37, no. 12, Dec. 1962, in Rutter Papers, UCSFA, MSS 94–54, box 5 of 13, Series II, file: School of Medicine—Exec Comm, 1971–72.

25. Kevles, *The Physicists*. Paul Starr, *The Social Transformation of American Medicine* (New York, 1982), 338–47. Nina Gilden Seavey, *A Paralyzing Fear: The Triumph over Polio in America* (New York, 1998). Patterson, *Grand Expectations*, 317–21.

26. *Time*, 25 July 1969, 16. Rickover, *New York Times*, 25 Oct. 1965, 39; Leonard, "The Last Word," *NYT Book Review*, 18 July 1971, 31. Leonard, "The Last Word: Should Science Be Shot?" *NYT Book Review*, 18 July 1971, 31, quoted in Daniel Kevles, *The Physicists*, 399–400.

27. Mario Savio, 3 Dec. 1964, Berkeley Web site: www.lib.berkeley.edu/MRC/saviotranscript

28. Abbie Hoffman, *Soon to Be Made into a Major Motion Picture* (Berkeley, 1980), 40. Roger Friedland, *Daily Californian*, 9 Dec. 1965, 1. *Stanford Daily*, 23 Sept. 1963, 1. Mordecai Briemberg, "Universities and Military—'End of Critical Intelligence,'" *Daily Californian*, 7 Dec. 1965, 10.

29. On Berkeley enrollments, Recorder of the Faculties, *University of California-Statistical Addenda* (UC Press, corresponding years). Joel Kugelmass, *Daily Californian*, 16 May 1966. Todd Gitlin, *The Sixties: Years of Hope, Days of Rage* (New York, 1987), 181.

30. Gitlin, *The Sixties: Years of Hope, Days of Rage*. W. J. Rorabaugh, *Berkeley at War; the 1960s* (New York, 1989), 144. "Earth Day/Week," 20 Apr. 1970, in BANC, file 7.9, Earth Day/Week. Dr. Marc Lappe, "Chemical and Biological Warfare; the Science of Public Death," October 1969, in BANC, File: 23:37 Student Research Facility, Pamphlet.

31. On science-activist groups, see BANC, 86/157c Social Protest Collection, esp. carton 7, Ecology/Electoral Politics, file: 7.3. 7.5, 7.10, 7.13 ("Would you eat DDT? You already do!" Friends of the Earth, 2 Dec. 1969), and 20:18. See also BANC, 66/157c Social Protest Collection, carton 2 Anti-Vietnam War, file 2.6. Patricia Gossel, Smithsonian Institute—NMAH, files: "Friends of the Earth," "Science for the People," and "Sheldon Krimsky."

32. Mark Schechner, "The University as Critic of Society—Where are We Left?" *Daily Californian*, 9 Nov. 1965, 7 and 10. Jacques Barzun, "Art and Science—How Soon the Fatal Dose?" *Daily Californian Weekly Magazine*, 11 Nov. 1965, 7. Nathan Glazer, "The University and Military Research: Glazer Replies," *Daily Californian*, 4 Jan. 1966, 13.

33. Jerry Rankin and Mary Lou McKinley, *Stanford Daily*, 13 Apr. 1960, 1. Jerry Rankin, *Stanford Daily*, 12 Jan. 1961, 1.

34. Ironically, while money poured into the biophysics laboratory, Stanford

announced that cost overruns prevented the university from keeping their earlier agreement with the San Francisco clinical hospital, as discussed in Chapter 4. *Stanford Daily,* 6 Apr. 1960, 1 and 6 July 1961, 2. *Stanford Daily,* 6 July 1961, 2. Leo Krulitz, *Stanford Daily,* 27 Nov. 1957, 1 and 12 Jan. 1959, 1. *Stanford Daily,* 5 Jan. 1960, 2.

35. *Stanford Daily,* 31 Oct. 1961, 1 and 1 Mar. 1962, 1. Mary Kay Becker and Robert Naylor, *Stanford Daily,* 23 May 1963. Elizabeth Freeman, *Stanford Daily,* 6 Mar. 1963, 1. Kirk Hanson, *Stanford Daily,* 19 Nov. 1965, 1; Justin Beck, *Stanford Daily,* 3 Feb. 1964, 1. *Stanford Daily,* 14 Jan. 1965, 4, including excerpts from Sterling's speech delivered on 13 Jan. 1965 to the Stanford Today and Tomorrow Convocation.

36. Graduate enrollments in the biosciences, 1964–65: biochemistry, 15%, biological sciences, 28%; metabolism and nutrition (food research), 35%; physiology, 34%. Undergraduate trends are often even more pronounced. For data on Stanford enrollment trends in the mid-1960s, see the Student University Bulletin, Directory of Officers and Students, by corresponding year for both graduate and undergraduate students. *Stanford Daily,* 23 Sept. 1963, 1; 25 May 1964, 4; 24 May 1964, 4; 26 Sept. 1963, 2; John Bonine, *Stanford Daily,* 4 Dec. 1964, 1. Jamie Hunter and David French, *Stanford Daily,* 16 Apr. 1963, 2. Kirk Hanson, *Stanford Daily,* 19 Nov. 1965, 1; Justin Beck, *Stanford Daily,* 3 Feb. 1964, 1.

37. Steve Baffrey, *Stanford Daily,* 13 Oct. 1959, 1; Jerry Rankin, *Stanford Daily,* 29 Oct. 1959, 1; and 5 Nov. 1959, 1. On the faculty revisiting the four-year program, see Office Memorandum regarding the Stanford Plan and its evaluation, 18 May 1965, in Lederberg Papers, SUA, SC 186, box 6, folder 2, Curriculum Committee, general, 1962–9.

38. Evidence of the strength of the movement were instances of anti-antiprotest, such as the math test on a student pamphlet: "Little Ho has 50 cents to buy bullets. He has 25 cents to buy band-aids. Some friends send him 15 cents for band-aids. Now, how much extra money does he have to buy bullets?" *Stanford Daily,* 1, 2, and 3 Nov. 1965, 1.

39. Tina Press, *Stanford Daily,* 28 Jan. 1966, 1. Bill Wertz, *Stanford Daily,* 3 Feb. 1966, 1. *Stanford Daily,* 4 Feb. 1966, 2. Dick Livingston, *Stanford Daily,* 26 Jan. 1966.

40. Jim Selna, *Stanford Daily,* 29 Mar. 1966, 1. John Buzan, *Stanford Daily,* 25 May 1965, 1.

41. *Synapse,* 15 Dec. 1965, 4.

42. *Synapse,* 5 Apr. 1965, 8 Oct. 1965, 2 Nov. 1965.

43. Roger Lang, "A Lesson Learned?" *Synapse,* 6 Dec. 1963, 2. Matsushima, *Synapse,* 7 Dec. 1964, 1. "Boat Rockers," *Synapse,* 8 Oct. 1965, 2.

44. Roger Lang, *Synapse,* 15 Nov. 1963, 2. *Synapse,* 31 May 1963; 8 Oct. 1965, 2; 6 Dec. 1963, 4.

45. Certainly the irony runs deep: a pinnacle of the humanist bioscience movement at UCSF turned out to be abortion. Frank Sarnquist, *Synapse,* 1 Dec. 1966, 1. *Redwood City Times Gazette,* 15 Oct. 1969, in Stanford Medical School Archives, 5360/4 S Medical Center News Bureau Scrapbooks, box 3, folder: June–Aug. 1970.

46. Marc Lappe, "Chemical and Biological Warfare: the Science of Public Death," in BANC, 86/157c, Social Protest collection, "The Right, cont.; Student Movement, Berkeley; Other," file 23:37, Student Research Facility. Allen Matusow, *The Unraveling of the Sixties,* 318–21.

47. "The Right, cont.; Student Movement, Berkeley; Other," in BANC, 86/

157c, Social Protest Collection, carton 23, file 23:37, Student Research Facility, especially Dr. Marc Lappe, "Chemical and Biological Warfare: The Science of Public Death." W. J. Rorabaugh, *Berkeley at War* (New York, 1989). Matusow, *The Unraveling of the Sixties*, 320.

48. Timeline of events from "Maggie's Farm; A Radical Guide To Stanford" booklet published by the Stanford Chapter of Students for a Democratic Society, in Pacific Studies Center, Mountain View, Calif., Archives: Stanford—Protest Movement, and in personal files. Interview transcripts of Lenny Siegel, director and curator of the Pacific Studies Center Archives.

49. *Stanford Review*, March—April 1966, in Pacific Studies Center, Mountain View, Calif., Archives: Stanford—Protest Movement.

50. "Maggie's Farm; A Radical Guide To Stanford," 38–39.

51. *Stanford Review*, March—April, 1966. Interview transcripts of Lenny Siegel.

52. Donald Worster, *Nature's Economy* (Cambridge, 1985).

Chapter 6: A Season of Policy Reform

Note to epigraph: George Daniels, "The Pure-Science Ideal and Democratic Culture," *Science*, 30 June 1967, 1699.

1. News and Comments, "NIH: Fountain Committee Issues Bitter Attack on Programs," *Science*, 3 Nov. 1967, 611.

2. FDA bill, submitted by Senator Kefauver on 12 Apr. 1961, designated S 1552; Barbara B. Troetel, "Three-Part Disharmony: The Transformation of the Food and Drug Administration in the 1970s" (Ph.D. diss., City University of New York, 1996); Theodore Brown, Jr., and Robert B. Allen, "The Progressive Populist," *Populists* (1996).

3. Daemmrich, "A Tale of Two Experts," *Society for the Social History of Medicine*, 152–57; Donna Hamilton, "1961: Spurring Drug Reforms to Prevent Birth Defects," *A History of the FDA* (FDA History Office: Nov. 1997), also on www.fda.gov; News and Comment, "The Administration of Federal Aid: A Monstrosity Has Been Created," *Science*, 7 July 1967, 43; Interview of Tom Kiley, legal counsel for Genentech Corporation, on 7 July 2000.

4. Troetel, "Three-Part Disharmony"; John F. Kennedy, "State of the Union—1962," *The Cumulated Indexes to the Public Papers of the Presidents of the United States, John F. Kennedy* (Millwood, 1977); *Stanford Daily*, 10 Feb. 1961 and 8 Mar. 1961.

5. News and Comment, "NIH: Fountain Committee Issues Bitter Attack on Programs," *Science*, 3 Nov. 1967, 611; Harold Orlans, "Developments in Federal Policy Toward University Research," *Science*, 10 Feb. 1967, 665.

6. Congressional Record, Senate, 17 May 1965, in BANC, CU-5, Series 5, box 14:23, folder: Secretary and Treasurer, file: Federal Legislation re: patents.

7. Allen Matusow, *The Unraveling of America* (New York, 1986), ch. 6; Judith Stein, *Running Steel, Running America* (Chapel Hill, 1998).

8. D. S. Greenberg, "Money for Research: LBJ's Advisers Urge Scientists to Seek Public Support," *Science*, 19 May 1967, 920.

9. Ibid., 920.

10. Ibid., 920–21.

11. Lyndon B. Johnson, "Special Message to Congress—1966," *The Cumulated Indexes to the Public Papers of the Presidents of the United States, Lyndon B. Johnson*

(Millwood, 1978); Julius Comroe, "Scientific Basis for Support of Biomedical Science," *Science*, 9 Apr. 1976, 105.

12. Comroe, "Scientific Basis for Support of Biomedical Science," 105; Greenberg, "Money for Research," 920–21; Hugh Davis Graham, *Civil Rights and the Presidency* (New York, 1992); Strickland, *Politics, Science and Dread Disease* (Cambridge, 1972), 209; Elizabeth Drew, "The Health Syndicate," *Atlantic Monthly* (Dec. 1967), 81.

13. Comroe, "Scientific Basis for Support of Biomedical Science," 105; "Project Hindsight," *Science* 18 Nov. and 2 Dec. 1966.

14. Strickland, *Politics, Science and Dread Disease*, 207; Joseph Newhouse to Dean Glaser, "9/8/69, L-16808, The RAND Corporation," in Lederberg Papers, SUA, SC 186, box 9, file: Medical School Administrative Comm., 1969; see also documents in the Rutter Papers, UCSF, MSS 94–54, box 6 of 13, series II, file: Council of Chairmen, 1973–74.

15. Philip Abelson, *Science*, 9 June 1967, editorial; Robert Semple, "President's Orders a Medical Review," *New York Times*, 28 June 1966, 35; Strickland, *Politics, Science and Dread Disease*, 207.

16. The theme of practical application began to resonate internally, too. Stanford physicians used *Life* magazine's popular special report on the "wonderously practical breakthroughs of modern science for all mankind" as a backdrop for a lecture series on "the far-out frontiers of medicine and biology"; an editor for *Science* turned against bioscientists, calling them "too conservative" and even postulated that the problems faced by all of the science disciplines might simply disappear if "American biologists began thinking of others" (*Life*, 59:11, 10 Sept. 1965); Roger Revelle, "International Biological Program," *Science*, 24 Feb. 1967, editorial; *New York Times*, 23 July 1965, 13; *Stanford Daily*, 8 Nov. 1965, 1, and 19 Oct. 1964 (Goldwater hammered away at federal waste, especially on a swing through Stanford on his way to the RNC convention in the Cow Palace in San Francisco where he would accept his party's nomination to run for the president of the United States in 1964); *Stanford Daily*, 17 Nov. 1966, 1 Dec. 1966, and 27 Oct. 66; Carl Djerassi, *The Pill, Pygmy Chimps, and Degas' Horse* (New York, 1992), 154; UCSF *Synapse*, 29 Apr. 1966, 9.

17. Strickland, *Politics, Science and Dread Disease*; Greenberg, "LBJ's Budget: Lean Fare Set Forth for Research and Development," *Science*, 27 Jan. 1967, 435; "Medical Research: NIH Wants Divorce from PHS," *Science*, 7 Apr. 1967, 45; Philip Abelson, "The Succession at NIH," *Science*, 28 Apr. 1967, editorial; Greenberg, "NIH Budget," *Science*, 26 May 1967, 1071.

18. Scientists had long ago isolated and purified a piece of Phi-X-174 DNA and found it exceedingly small—containing enough information to create only five or six total genes—and that its very special trait—it had a circular shape—made it especially conducive to DNA synthesis experiments. Believing—or guessing, as Kornberg would later admit—that the shape rather than the order of the bases might make the DNA biologically active, Kornberg's research team tried endless experiments to join or ligate the two ends together to create a circle. Looking back, two problems proved especially challenging. First, even the simplest strand of DNA is incredibly complex—a few simple bacteria had hundreds of genes, most had thousands—which makes the manipulation of DNA quite difficult. The second, and greater obstacle of the two, is finding the enzymatic mechanism that can bring "life" to a synthetic piece of DNA (Arthur Kornberg, *For the Love of Enzymes* [Cambridge, 1989], ch. 6).

19. Greenberg, News & Comments, "The Synthesis of DNA: How They Spread the Good News," *Science*, 22 Dec. 1967, 1548.

20. Joshua Lederberg, "'Creation of Life' Is More Than a Description," *Washington Post,* 23 Dec. 1967, in Sally Hughes, "Arthur Kornberg, M.D.: Biochemistry at Stanford, Biotechnology at DNAX" (Berkeley, 1998), appendix F; see also Mark Weinberger, "Synthesis of DNA 'No Breakthrough,'" *Stanford Daily,* 3 Jan. 1968, 1; Greenberg, "The Synthesis of DNA," 1548; Maxine Singer, "In Vitro Synthesis of DNA," *Science,* 23 Dec. 1967, 1550; Arthur Kornberg, "Biochemist Arthur Kornberg: A Lifelong Love Affair With Enzymes," *The Scientist,* 3[17]:13 (4 Sept. 1989); Kornberg, *For the Love of Enzymes,* 192–206 and 202.

21. Curiously, Kornberg's wish to call his own scientific achievement something other than "creating life in the test tube" is somewhat disingenuous considering that he uses that exact same phrase as the title for chapter 6 in his own book (Kornberg, *For the Love of Enzymes,* 202–3).

22. Kornberg, *For the Love of Enzymes,* 202–4.

23. Bryce Nelson, "Space Budget: Congress Is In a Critical, Cutting Mood," *Science,* 14 July 1967, 171; Philip Boffey, "Federal Research Funds," *Science,* 8 Dec. 1967, 1286; Kevles, *The Physicists,* 397.

24. Abelson, "The Succession at NIH," *Science* (9 June 1967), editorial; Abelson, "A Partisan Attack on Research," *Science,* 9 June 1967, editorial; Senator Harris, National Commission on Health Science and Society, "Hearings Before the Subcommittee on Governmental Research" (Washington, D.C., 1968), Ninetieth Congress, Second Session, on S.J. Res. 145 (7 and 8 Mar. 1968), 2; on Kennedy and Humphrey, see Nelson, "Space Budget," 171; "NIH: Fountain Committee Issues Bitter Attack on Programs," *Science,* 13 Nov. 1967, 611.

25. A dominant coalition of moderate Republicans and Southern Democrats moved first, agreeing to reduce federal spending reduced as their price for supporting President Johnson's request for a 10-percent war tax surcharge. Nelson, "Space Budget," 171; Greenberg, "Federal Economizing: House Votes to Take It Out of R & D," *Science,* 27 Oct. 1967, 474; Elizabeth Drew, "The Health Syndicate," *Atlantic Monthly,* 19 Dec. 1967, 78.

26. "Administration of Research Grants in the Public Health Service" (Washington, D.C., 1967), also known as the "Fountain Report" (D-NC); News & Comments, "NIH: Fountain Committee Issues Bitter Attack on Programs," 611.

27. National Commission on Health Science and Society, "Hearings Before the Subcommittee on Governmental Research" (Washington, D.C., 1968), Ninetieth Congress, Second Session, on S.J. Res. 145 (7 and 8 Mar. 1968).

28. National Commission on Health Science and Society, "Hearings Before the Subcommittee on Governmental Research," S.J. Res. 145 (6 and 7 Mar. 1968).

29. National Commission on Health Science and Society, "Hearings Before the Subcommittee on Governmental Research" (7 Mar. 1968), 39–48.

30. National Commission on Health Science and Society, "Hearings Before the Subcommittee on Governmental Research" (8 Mar. 1968), 49–53.

31. National Commission on Health Science and Society, "Hearings Before the Subcommittee on Governmental Research" (8 Mar. 1968), 54–69; Kornberg Papers, SUA, SC 359, 89–063, box 1, file: "Correspondence; 1968."

32. Greenberg, "Federal Economizing," 473.

33. Sherman to Gubser, 22 Dec. 1967, in Kornberg's Papers, SUA SC 359, 89–063, box 31, file: Political 1961–4.

34. Greenberg, "LBJ's Budget," 435.

35. Ibid.

36. Michael Sweeney, *Stanford Daily,* 26 Jan. 1968, 1; Ketcham to Vice Chan-

cellor Sammet, 22 Nov. 1967, President's Papers, BANC, CU-5, series 5, box 143:23, file: Grants and Contracts.

37. Greenberg, "LBJ's Budget," 434.

38. Edwards to chancellors of the UC, 22 Nov. 1966, President's Papers, BANC, CU-5, series 5, box 143:23, file: Grants and Contracts.

39. Ferber to President Wellman, 8 Aug. 1967, President's Papers, BANC, CU-5, series 5, box 143:23, file: Grants and Contracts.

40. Boffey, "Federal Research Funds: Science Caught in Budget Squeeze," *Science*, 8 Dec. 1967, 1288.

41. "Themis: DOD Plan to Spread Wealth Raises Questions in Academe," *Science*, 9 June 1967, 48.

42. Ferber to President Wellman, 8 Aug. 1967, President's Papers, BANC, CU-5, series 5, box 143:23, file: Grants and Contracts; Graph, *Science*, 1 Sept. 1967, 991.

43. "Special Issue—Hospitals OR Homes for the Haight," *Haight Action*, 15 (Aug. 1971), in Department of the History of Health Sciences Papers, UCSF, AR 87–46, file: Neighborhood PR 1964–74 (1), 1971; Vernon Sturgeon Oral History R-25, "State Senator, Reagan Advisor, and PUC Commissioner, 1960–1974," www.ss.ca.gov/archives/level3_ohguide4s.html; BANC, CU-5, series 5, box 156:18, file: California State Legislature—Gen., June 1964–June 1966.

44. Kidner to President Kerr, 17 Aug. 1965, in BANC, CU-5, series 5, box 156:18, file: California State Legislature—General, June 1964—June 1966, 8/17/65, with attached transcripts of Speaker Unruh "Dunderbeck's Machine," Lake Arrowhead (7 July 1965), where Unruh declared "education is the greatest enemy of the Legislature."

45. Countless jabs at Governor Reagan appear in Berkeley and UCSF student newspapers. For instance, see Apple's parody "Ronnie for Governor," *Synapse*, 10 Mar. 1965, 4: "Ronnie Reagan is his name, A ketchup bottle his claim to fame / Hold it up in your RIGHT hand; Play 'America' throughout the land. / Crack a whip over strikers' backs; Let's outlaw the income tax: / Taxes pay the Commie scholastic, To spook John Birch with smirch sarcastic. / To end our war in perfect calm, Let's pave over all Vietnam. / Run for governor the Hollywood mob, It's 'earn while you learn' the state's top job. / Ronnie, my boy, with fame you're delirious, but the governor's job is really serious; / It's the state of California you're asking to lead, / But Ronnie, didn't you know: there's no script to read?" SCR—42 filed with the Sec. of State, 11 Aug. 1967, and Richard Wolfe to VP Bolton, 9 Oct. 1967, all in President's Papers, BANC, CU-5, series 5, box 143:23, file: Grants and Contracts; Duesberg to Radiation Safety Committee, 26 Dec. 1974, in Stent Papers, BANC MSS 99/149z, carton 34, file: 34:15, Molecular Biology, Correspondence, 1965–75; *San Jose Mercury News*, 23 Apr. 1968, in SUA, 5360/4 S Medical Center News Bureau Scrapbook, box 1, Jan. 59–Spring 67.

46. Annonymous handwritten note, 15 Jan. 1973, attached to *Stanford M.D.*, summer/fall 1971, vol. 10, no. 3, in SUA, SC 358, box 17, file: Binder Faculty Meetings, 1971–3; John Cooper, M.D. "The Carnegie Commission and the Academic Medical Center," Stanford Medical Alumni Association, 6–10, cut-out article in Kornberg Papers, SC 359, Kornberg, 89–063, box 31, file: Correspondence, "Political 1965–8"; Natalie D. Spingarn, *Heartbeat: The Politics of Health Research* (New York, 1976), 9.

47. Strickland, *Politics, Science, and Dread Disease*, 214 and 217.

48. Cooper to the Assembly of the AAMC, 21 October 1974, in Rutter Papers, UCSF Archives, MSS 94–54, box 6 of 13, series II, file: Council of Chairmen,

1973–4; Minutes of the Advisory Board of the School of Medicine, 23 Sept. 1971, in Rutter Papers, MSS 94–54, box 6 of 13, series II, file: School of Medicine—Advisory Comm, 1969–71.

49. Greenberg, "LBJ's Budget," 434; Saunders to Senator Hill, 11 Mar. 1964, in CU-5, series 5, box 96:17; Minutes of the Advisory Board of the School of Medicine, 2 Feb. 1972, in Rutter Papers, MSS 94–54, UCSF, box 6 of 13, series II, file: School of Medicine—Advisory Comm, 1972–5; Strickland, *Politics, Science and Dread Disease*, 219 and 220.

50. Belanger, *Enabling American Innovation*; Tony Appel, *Shaping Biology*, ch. 9.

51. Notes on Casey Bill, AB-5, 31 10 66, in BANC, CU-5, series 5, box 23:8, file: Budget—Hospital Subsidies, General.

52. Jeannie Rosoff to Dr. Rutter, 17 July 1970, in Rutter Papers, MSS 94–54, Rutter papers, box 5 of 13, series II, file: School of Medicine—Exec. Comm., 1970–71; Minutes of the Advisory Board, Dean's Announcements, 23 Dec. 1970, in Rutter Papers, MSS 94–54, box 5 of 13, file: School of Medicine Executive committee, 1969–70.

53. Jardetsky to Rich, 10 Dec. 1973, in Lederberg Papers, SUA, SC 186, box 1, file: Basic Science Planning Committee; Morales to Chancellor Fleming, 13 May 1968, regarding Academic Planning Comm. Meeting, 16 Apr. 1968, in UCSF, AR 86–32, carton 1, file: 2, 20 and 26, esp. "Fifth School Proposal" and carton 1, file: Human Biology, 1968; Dean Fretter's notes, 28 Oct. 1963, and Fretter to Chancellor Strong, 5 Nov. 1963, and Fretter to Barker, 18 Mar. 1964, all in BANC, CU-149, 662 T, 75:11 Biochem. Again, there are countless examples of the divide between practical and pure research, such as the program on the muscular dystrophy in chickens, transferred from Berkeley to the veterinary program at UC Davis (*Daily Californian*, 18 Oct. 1964, 6).

54. Berg to graduate students, 26 Sept. 1973, in Kornberg Papers, SUA, SC 358, box 17, file: Faculty Meetings, 1971–3; Hirschler to Stent, 9 Dec. 1969, in Stent Papers, BANC, MSS 99/149z, Carton 34, file: 34:15, Molecular Biology: Correspondence, 1965–75; Michael Sweeney, *Stanford Daily*, 26 Jan. 1968, 5; Vettel, *A Scientific Maverick; An Oral History with Donald Glaser* (Berkeley, 2005); Sally Smith Hughes, "William J. Rutter, Ph.D.: The Department of Biochemistry and the Molecular Approach to Biomedicine" (San Francisco, 1998), 69; Pat Weir to Stent, 30 Oct. 1969, Stent to "Barbara," 14 Aug. 1970, and Wofsy to Lawrence Sullivan, 27 Apr. 1970, all in Chancellor Papers, BANC, CU-149, box 115, file: 38, Bacteriology & Immunology, L & S.

55. Gilmartin to Cuthbertson, 4 Apr. 1969, in Lyman Papers, SUA, SC99, box 14, file: Humanities and Sciences Hopkins Marine Station; "Higher Education: Scrambling for the Philanthropic Dollar," *Science*, 28 Apr. 1967, 494; Henry Alley, "SU May Buy Out Split Hospital," *Stanford Daily*, 31 Jan. 1966, 1; Sally Hughes, "Making Dollars Out of DNA," *Isis*, 2001, 92: 545–46.

56. Cunningham to Kerr, 29 Jan. 1963, in BANC, CU-5, series 5, box 143:5, file: Campus Gifts and Endowments Development Program—San Francisco.

57. The report lists alternative fund-raising strategies, including a letter-writing campaign, to those "affected by serious illness, [which plays on] an emotional motivation which spurs the desire to contribute," Cunningham to Kerr, 29 Jan. 1963, in BANC, CU-5, series 5, box 14, file: the Board of Regents; Comm. on Research; Wellman to Kerr, 11 Mar. 1964, and Saunders to Kerr, 5 May 1966, and Cunningham to Wellman, 17 Feb. 1964, all in BANC, CU-5, series 5, box 14, file: Campus Gifts and Endowments Development Program—San Francisco.

Chapter 7. Crossing the Threshold

Note to epigraph: Max Planck, *Scientific Autobiography and Other Papers,* trans.- Frank Gaynor, pp. 33–34 (1950).

1. Paul Aebersold, *Biotechnology Backstage* (Rockville, 1998).

2. "Control of Life," *Life,* 59:11 (10 Sept. 1965), 59; Subcommittee for a General Curricular Pathway, memo, 20 Apr. 1967, in SUA, Kornberg Papers, SC 359, box 1, file: Biochemistry Department, Faculty Minutes, 1967; Martin Cline to David Wood, 8 June 1967, in UCSF, S/M, AR 90–56, Dean's Office Records, 1936–89, carton 5 of 13, file: Organized Research—Cancer Research Institute, 1966–67; "The Organization Problem," in BANC, unlabeled DIB, file: 34:41 Biology I, Correspondence 1966–7.

3. *Medical Center Memo,* in SUMC Archives, 5330/4, Med Center News Bureau Scrapbooks, boxes 1967–68; *The Healing Arts, A Report to the Community from the Stanford University Medical Center,* in SUMC Archives, 5330/4 H, boxes 1971–73; Subcomm for a Gen Curricular Pathway to the Comm on Medical Educ, memo and notes from 20 Apr. 1967 and 17 May 1967, in SUA, Kornberg Papers, SC 359, box 1, file: "Biochemistry Dept., Faculty Minutes, 1967."

4. Stanford took advantage of new cancer research policies too. Medical microbiologist Carlton Schwerdt studied cancer viruses in rabbits; David Kalman and Goldstein explored biochemical and metabolic reactions of cancerous tumors; Edward Mansour conducted research on the cellular metabolism of cancer; Philip Hanawalt investigated the repair of DNA damaged by cancer; Lederberg traced the genetic mechanisms of inherited cancer; Herzenberg explored the evolution of normal and cancerous cells grown in culture; and Kaplan conducted radiation carcinogenesis in mice. (SUA, SC 359, Kornberg, 89–063, box 31, file: Stanford Correspondence, 1961–66, 1964, and box 26, file: Stanford Correspondence 1967–8; SUA, 5360/4 S Medical Center News Bureau Scrapbooks). Robert Proctor, *Cancer Wars* (New York, 1995); James Patterson, *The Dread Disease: Cancer and Modern American Culture* (Cambridge, 1987); Nancy Rockafellar, "Interviews with Rudi Schmid" (San Francisco, 1998), 96–97; Sally Hughes, "John Lawrence: Nuclear Medicine Pioneer and Director of Donner Laboratory" (Berkeley, 2000); Rita Carroll, "Conversations with Dr. Leslie 'Latty' Bennett" (San Francisco, 1992); Hughes, "Paul Berg: A Stanford Professor's Career in Biochemistry, Science Politics, and the Biotechnology Industry" (Berkeley, 1997); Wood to Cullen, Dean of Medicine, 24 Apr. 68, in UCSF Archives, AR 90–56, S/M, Dean's Office Record's, 1936–89, carton 5 of 13, file: Organized Research—Cancer Research Institute, 1968–69 and 1970–71.

5. On "massaging" soft money, see virtually any oral history of a UCSF investigator, including Nancy Rockafellar, "Interviews with W. F. Ganong: Neuroendocrinology in the Academic Medical School" (San Francisco, 1995), 35; Rockafellar, "Interviews with John Clements: The Story of Pulmonary Surfactant and Basic Science in the CVRI" (San Francisco, 1998), 30.

6. Victor Twitty to President Sterling, 29 Nov. 1955, in SUA, Terman Papers, box 3, folder 4, file 7, and Terman notes, 24 Apr. 57, folder 14, file 2; Martha Maskall, "Biology's New Core Curriculum Plans," *Stanford Daily,* 31 May 1963; RAC, collection RF, record group 1.2, series 200A, box 148, folder 1331, file: Grad Stud Pam.

7. Other faculty in the biology department tempted by the promises of basic research were Donald Abbott, Richard Holm, Robert Page, and David Regnery. Maskall, "Biology Department Reveals New Core Curriculum Plans," *Stanford Daily,* 31 May 1963.

8. Maskall, *Stanford Daily*, 31 May 1963.

9. Maskall, *Stanford Daily*, 31 May 1963.

10. Grobstein, *Stanford Daily*, 14 Jan. 1964; Kornberg to Grobstein and Terman, 16 Oct. and 28 Oct. 1963, in SUA, SC 359, Kornberg, 89–063, box 31, file: Stanford Correspondence, 1961–66.

11. Grobstein, *Stanford Daily*, 14 Jan. 1964, 2.

12. Grobstein, *Stanford Daily*, 14 and 16 Jan. 1964; *Stanford Daily*, 2 Dec 1964, 1.

13. Manuel Morales to Dean William Reinhardt, 26 June 1963, in UCSF Archives, Dean's Office Papers, AR 90–56, carton 1, file 17, Biochemistry 1961–6.

14. Emil Smith to Dean William Reinhardt, 6 Mar. 1964, in UCSF Archives, Dean's Office, AR 90–56, carton 12, file 24–2, Biochemistry Search Committee; "Department Report and Roster," in UCSF Archives, AR 90–56, carton 1, file 16, Biochemistry 1966.

15. Dean Krevans Report to the Exec. Comm., Faculty Council and Council of Chairmen, in UCSF Archives, MSS 94–54, Rutter Papers, box 6 of 13, series II, file: Council of Chairmen, 1973–4.

16. Jawetz to Krevans, 12 Oct. 1971, and Dean Krevans to Jawetz, 29 Nov. 1971, all in UCSF Archives, document section 42, carton 2, file: Departments—Microbiology, 1968–73; Morales to Dean Cullen, 28 Feb. 1967, and Walter Stoeckenius to Dean Cullen, 1 May 1968, in Dean's Office Papers, AR 90–56, carton 12, file 24–2, Biochemistry Search Committee; Morales to Dean Reinhardt, 21 Oct. 1963, in UCSF Archives, Dean's Office Papers, AR 90–56, carton 12, file 24–2, Biochemistry Search Committee; Morales to Dean William Reinhardt, 26 June 1963, UCSF Archives, Dean's Office Papers, AR 90–56, carton 1, file 17, Biochemistry 1961–6.

17. Hughes, "William J. Rutter, Ph.D.: The Department of Biochemistry and the Molecular Approach to Biomedicine,"(Berkeley, 1998), 18; Hughes, "Interviews with Herbert Boyer" (Berkeley, 2001).

18. Hughes, "William J. Rutter"; Dean Cullen to Rutter, 12 July 1967, in UCSF Archives, the Dean's Office, AR 90–56, carton 1, file 16 and 17, Biochemistry 1961–6.

19. Julia Bazar and Robin Chandler, "The William Rutter Papers—Finding Aid," UCSF Archives, 8/95; *Alumni-Faculty Association Bulletin*, vol. 22, no. 1, spring 1978, 2; Rutter to Members of the Exec. Comm., 17 Feb. 1970, in UCSF Archives, MSS-54, Rutter Papers, box 5 of 13, file: School of Medicine Executive Committee, 1969–70; Hughes, "William J. Rutter"; Hughes, "Interviews with Herbert Boyer."

20. The three Nobel Prize winners at UCSF are: J. Michael Bishop, Harold Varmus, and Stanley Prusiner. Even UCSF's administration departed from their past timidity and agreed to change the name of the biochemistry department to "Biochemistry and Biophysics" (Hughes, "William J. Rutter"and "Interviews with Herbert Boyer"); see also reference to NIH Training Grant AI 00299, in UCSF archives, carton 2, file: Dept—Microbiology, 1968–73.

21. Advisory Committee on Inter-Campus Medical Teaching and Research to Sproul, 2 Oct. 1950, and Johnson to Sproul, 26 Oct. 1951, Kerr to Sproul, 13 Mar. 1958, all in BANC, Department of the History of Health Sciences, AR 87–46, file: Institute of Experimental Biology.

22. Li, "Isolation and Properties of a New, Biologically Active Peptide from Sheep Pituitary Glands," *Journal of Biological Chemistry*, Apr. 1964, and "Lipotro-

pin," *Nature*, Feb. 1964 (vol. 201), 924. Sanger to Li, 8 Oct. 1956, and Bauner et al., 12 Oct. 1964, all in UCSF Archives, Li Papers, MSS 88–89, series I: corresp Sub-Series 2, subject, carton 17 of 56, file: MSH Controversy Corresp 1956–7, Lipotropin Challenges, 1964–5. Kerr to Pres Sproul, 13 Mar. 1958, in BANC, Dept of the Hist of Health Sci, AR 87–46, file: Institute of Experimental Biology.

23. Berkeley staff continued to criticize Li's work through the 1970s, though one scholar privately admitted to a reporter that synthetic HGH "certainly warrants further investigation . . . but for God's sake don't mention my name" (Edward Edelson, *Redbook Magazine*, Nov. 1972, 86). "A Tough Four Years," and "Major UC Research Feat—An Artificial Hormone," *San Francisco Chronicle*, 7 Jan. 1971, in UCSF Archives, Li Papers, MSS 88–89, Series I: corresp sub-series 2, carton 17 of 56, file: MSH Controversy Correspondence 1956–57, Lipotropin Challenges, 1964–65; Rutter Papers, UCSF Archives, MSS, files: HGH in "general correspondence."

24. Paul Rabinow, *Making PCR* (Chicago, 1996), 37–45; Patricia Gossel notes of interview with Mertz, 2 July 1992, and AMWS, 17th edition, 1989/90, 343, all in Gossel's files, Smithsonian Institution, NMAH; Kornberg, *For the Love of Enzymes*; Susan Wright, "Recombinant DNA Technology and Its Social Transformation," *Osiris*, 2nd Series, 1986, 2: 310; Hughes, "Paul Berg: A Stanford Professor's Career in Biochemistry"; interview notes with Tom Kiley, 7 July 2000; Hughes, "Interview with Mary Betlach," unpublished oral history (Berkeley, 1994).

25. Hughes, "Arthur Kornberg" (Berkeley, 1998), 136.

26. There are a number of sources that discuss anti-Semitism in the medical schools. As for anti-Semitism and how it related to the biosciences in the Bay Area universities, see Sally Hughes, "Arthur Kornberg," 181; Hughes, "Paul Berg."

27. Rita Carroll, "Conversations with Dr. Leslie 'Latty' Bennett," 26; Wendell Stanley Papers, BANC, box 8, file 61: General Correspondence, Franklin—Stanley.

28. Margaret Rossiter, *Women Scientists in America: Before Affirmative Action, 1940–1972* (Baltimore, 1995); Maresi Nerad has shown a connection between gender and departmental status at Berkeley during the 1950s; quantitative and qualitative evidence drawn from the biosciences extends her conclusions, suggesting a direct correlation between women and their appointment to so-considered less prestigious practical research programs (Maresi Nerad, *The Academic Kitchen* [Albany, 1999]).

29. Nancy Rockafellar, "Interviews with W. F. Ganong" (San Francisco, 1998), 27–28; interview notes with Tom Kiley, 7 July 2000.

30. A comparison of students born outside the United States with native-born students who enrolled in Stanford's bioscience programs supports this conclusion: for instance, between 1951–54 there were 2 foreign-born students, but from 1967–70 there were 20 (Stanford Annual Commence Records list place of birth, and SUA, SC 358, Berg Papers, box 17, binder: Department of Biochemistry, 1973–, file: faculty meetings, 1971–1973, and file: Department of Biochemistry, 1973/4.) For another example, during approximately the same time period, foreign-born graduate students in the CVRI at UCSF almost doubled, from 10 to 20 (Comroe to Saunders, 5 Mar. 1963, in UCSF Archives, Saunders' Papers, unsorted files, MSS 90–73, file: S/M, 1963; SUA, Lederberg Papers, SC 186, box 10, folder 7, Study of Med Educ at Stanford [as a comparison with UCSF], file: Med Educ in the US, 1970–1; see also *JAMA*, 22 Nov. 1971, vol. 218).

31. Margaret Rossiter, *Women Scientists in America.* There is evidence of a double standard. As for qualitative evidence, faculty teaching biochemistry at Stanford identified female students in pure research as the lowest performers, which seems highly unlikely because women in research had more experience and training in biochemistry than men who were enrolled as medical students. As for quantitative evidence of gender discrimination, there are numerous examples of female scientists and male medical students taking the same bioscience courses, and yet men consistently receiving higher marks. Kornberg Papers, SUA, SC 359, Kornberg 89–063, Biochemistry Faculty Minutes, 18 Dec. 1967, and Biochemistry Faculty Minutes, 23 Feb. 1968.

32. *Synapse,* 29 Oct. 1962; Rabinow, *Making PCR,* 106 (interview of Ellen Daniell).

33. Walter Blum, "Why Women Become Doctors," *San Francisco Sunday Examiner and Chronicle; California Living,* 30 Mar. 1969, 34; Stanford Annual Commencement Records, Stanford 1967, Order of Exercises; Student University Bulletin, Autumn qtr. 1967, and by corresponding year (published by the university). University of California Statistical Addenda, Compiled by the Recorder of the Faculties, 1919–47 (University of California Press), later listed as the University of California, Statistical Summary, Students and Faculty, 1967; Comroe to Saunders, 5 Mar. 1963, in UCSF Archives, Saunders' Papers, unsorted files, MSS 90–73, file: S/M, 1963, Long-range department goals. University of California, Announcements of the Medical School (University of California Press), 1967 and 1968.

34. Further evidence of a protest strategy that worked at crosspurposes: the AWIS struggled for equal pay while the NOW subchapter sought equal representation (Margaret Rossiter, *Women Scientists in America,* 380 and 369, respectively).

35. Offensive sexual remarks were certainly not limited to Stanford or UCSF. For instance, throughout the 1950s, a Berkeley laboratory ran a job advertisement in the employment section of the classified ads that sought "a girl to fill out forms" (David Kaiser, MIT, "The Postwar Suburbanization of American Physics," unpublished conference paper for panel "Making and Educating People in Cold War America," at the 2002 OAH Annual Meeting, 12 Apr. 2002); *Stanford Daily,* 9 Mar. 1967, guest editorial; *Synapse,* 22 Oct. 1971, 2, and 1 Dec. 1966, 2; Rossiter, *Women Scientists in America,* 370. On the surge of women enrolled in the biosciences in the Bay Area, see Recorder of the Faculties, *University of California—Statistical Addenda* (University of California Press, corresponding years); Stanford University, *University Directory* (corresponding years). A few feminist historians address this remarkable transformation. In the 1950s and early 1960s, bioscientists were overwhelmingly men and considered the body—particularly, the female body—much like the way they viewed applied research: both were forbidden research topics. By the late 1960s, women overcame to a significant degree their "socially imposed silence" and sought access to the biosciences because this was a field in which the body was, by practice, the central research question (Alice Wexler, *Mapping Fate* [Berkeley, 1995], introduction; see also Evelyn Fox Keller and Thomas Laqueur, *Making Sex: Body and Gender from the Greeks to Freud* [Cambridge, 1990]). There are a number of analytical problems with this argument: for one, bioscience research and people's attitudes about it remain inherently static, while the only change is the increase in the number of women in bioscience research.

36. Rossiter, *Women Scientists in America* ch. 16. Quantitative evidence drawn from Bay Area university bioscience programs supports Rossiter's arguments: for

instance, at UCSF, see Jawetz to Bennet, 7 Feb. 1974, summary report on the Department of Microbiology, 1968–73, and William Rutter, 1973 Research Report, all in UCSF Archives, MSS document section 42, carton 2, file: Departments—Microbiology, 1953–63, and 1968–73; *Synapse*, 8 Oct. 1971, 4.

37. This account of the race to develop recombinant DNA and cloning techniques in Bay Area university bioscience laboratories is not a comprehensive study of the rise of bioengineering; a number of fine, in-depth studies already exist that trace its elegant evolution. Instead, this portrait focuses on certain critical discoveries in local laboratories and gives considerable attention to personalities and motivations in bioscience research. Focusing on the culture of science and scientific wrong turns captures a young, impatient, dynamic field driving toward new bioscience frontiers and highlights many of the organizational and demographic changes discussed earlier in this chapter. Perhaps the most comprehensive account of the race to synthesize new and improved genes is Stephen Hall, *Invisible Frontiers* (New York, 1987); it is an internal story about the first recombinant DNA experiment.

38. E. L. Tatum predicted "bioengineering" in the late 1950s, but investigators kept this line of experimentation in check by emphasizing pure research. Hughes, "Paul Berg, 62, 147; Susan Wright, "Recombinant DNA Technologies," 2, 306.

39. Despite his early objections, Kornberg eventually accepted the fundamental value of Berg's applied interests: Berg was "really interested in trying to do something new, something different, that other people were not doing; it is natural to ask if gene expressions that work in simple bacteria are the same that work in eukaryotic cells." However, at the time Kornberg believed Berg would "waste his talents" (Hughes, "Paul Berg," 52).

40. Hughes, "Paul Berg," 59.

41. Joshua Lederberg, "Genetics to Bacteria," grant application to the National Institutes of Health, No. AI 05160–11, December 20, 1967; Wright, "Recombinant DNA Technologies," 310.

42. Stanley Cohen, "Biotechnology at 25: Perspectives on History, Science and Society," lecture given 13 Mar. 1999; audio and text-based transcripts obtained through the Bancroft Library biotech Web site: http://bancroft.berkeley.edu/Biotech/symponsium/cohen/text.html.

43. Hughes, "Paul Berg," 185.

44. Hughes, "Paul Berg," 132.

45. Hughes, "Paul Berg," 126, 134; and Hughes, "Arthur Kornberg," 146; and Kornberg, *For the Love of Enzymes* (Cambridge, 1989), 276.

46. Hughes, "Paul Berg," 102.

47. Ibid., 53, 59, 75, 100, 147, and 162.

48. Hughes, "William J. Rutter," i, 9, 87, and 92.

49. Hughes, "Herbert Boyer," introduction, 1; Wright, "Recombinant DNA Technologies," 312; Hughes, "Paul Berg," 120; Kornberg, *For the Love of Enzymes*, 281; and Hughes, "William J. Rutter," 86.

50. Kornberg, *For the Love of Enzymes*, 276; Cohen, "Biotechnology at 25," lecture transcripts found on BANC biotech Web site.

51. Arthur Kornberg, *For the Love of Enzymes*, 276.

52. Hughes, "Paul Berg," 76, 102, 104, and 126; Wright, "Recombinant DNA Technologies," 311.

53. Kornberg, *For the Love of Enzymes*, 280–81.

54. Kornberg, *For the Love of Enzymes*, 128; AMWS 17th ed., 1989–90, 343, and

Gossel's interview with Mertz, 2 July 1992, at SI-NMAH, Div of Sci and Med; Hughes, "Paul Berg," 119.

55. Hughes, "Paul Berg," 100; Hughes, "Interviews with Mary Betlach," unpub. and unfinished trans., 1 and 7.

56. Kornberg, *For the Love of Enzymes,* 140–45; Hughes, "Herbert Boyer," 32; Helling to McNiff, 11 June 1992, "Science in American Life," and Gossel's interview with Mertz, 2 July 1992; Hughes, "Paul Berg," 100 and 123.

57. Jackson, Symons, and Berg, "Biochemical Method for Inserting New Genetic Information into DNA of Simian Virus 40," *PNAS,* 1972, 69; Hughes," 133.

58. Hughes, "Paul Berg," 148; Cohen, "Biotechnology at 25."

59. Hughes, "Paul Berg," 110–11.

60. Wright, "Recombinant DNA Technologies," 310.

61. Stephen Hall, *Invisible Frontiers* (New York, 1987) , 8; Wright, "Recombinant DNA Technologies," 312.

62. Hughes, "Paul Berg," 109 and 112.

63. Lobban's experience in academia is particularly instructive. Lobban remembers the competitive "nightmare" to conduct applied research, as well as the provincial indifference of others: "Not a single person . . . understood the implications of being able to join DNA fragments together at will, let alone found it glamorous or even mildly interesting. If I got any reaction besides bemused silence, it typically took the form of a dismissal." One year later, Lobban left the biosciences and entered Stanford's graduate program in electrical engineering; Mertz accepted a research position with a private company. Kornberg, *For the Love of Enzymes,* 141–46; Hughes, "Paul Berg," 123, 134, 139, 146, and 161; see also Berg to Koprowski, 22 May 1973, in Berg Papers, SUA, SC358, box 3, folder: 1973; Hughes, "Interviews with Mary Betlach," 48; Patricia Gossel's files on Mertz at the SI—NMAH, Div of Sci and Med; Rabinow, *Making PCR,* 45, from interview of Gelfand.

64. Hughes, "Paul Berg," 38, 68, 91–93, 111–13 and 148; Hughes, "Interviews with Mary Betlach," 30–31; Hughes, "Paul Berg," 11 and 80.

Chapter 8. Cetus

1. Miller memo, 7 Feb. 1962, BANC, CU-149, Box 75:33, file: Molecular Biology, proposed.

2. Vettel, "*Conversations with a Maverick Scientist*"; Martin C. Easter, Rapid Microbiological Methods in the Pharmaceutical Industry (Weimar, TX, 2003).

3. Among others, see Geigert, Hansen, McDowell, Merrill, and Ward, "Microbiological Assay Utilizing an Automatic Zone Scanner," *Developments in Industrial Microbiology,* vol. 17, ch. 14.

4. Farley to all partners, memo 1, 18 Nov. 1971, box C0120495971, unsorted files, and box C0120565135, file: scanners, "Computer Identification of Bacteria by Colony Morphology"; Geigert et al., "Microbiological Assay Utilizing an Automatic Zone Scanner," all in Chiron archives, Cetus papers (hereafter, "CACP").

5. "Toward Automated Microbiology," *Laboratory Management,* July 1971, 24–26. "Automating the Bac't Lab; Technicon's TAAS," in CACP; "A Closer View of Antibiotics," *Medical World News,* 8 Oct. 1971, 49.

6. Sally Smith Hughes, "Paul Berg: A Stanford Professor's Career in Bio-

chemistry, Science, Politics, and the Biotechnology Industry" (Berkeley, 1997), 52.

7. Sally Smith Hughes, *Moshe Alafi Oral History Interview* (Berkeley, 2004).

8. Martin Kenney and Richard Florida, "Venture Capital in Silicon Valley: Fueling New Firm Formation," in Martin Kenney, ed., *Understanding Silicon Valley: the Anatomy of an Entrepreneurial Region* (Palo Alto, 2000), 98–123; Wilson, *The New Venturers* (Menlo Park, 1985), 21–24.

9. Moshe Alafi, private papers, held by ROHO.

10. Neither Glaser nor Alafi remember the precise moment when they first met—or do they seem to care.

11. Hughes, *Moshe Alafi Oral History*, 30.

12. Wattenburg and Donald Glaser, General Partners, "Business Plan," Donald Glaser Papers, unsorted, BANC, carton 6–5, file: Berkeley Scientific Laboratories, 1965–.

13. Wattenburg to Glaser, handwritten, sometime in late 1965, and Wattenburg to Connick, 4 Apr. 1966, all in Glaser papers unsorted, BANC, carton 6–5, file: Berkeley Scientific Laboratories, 1965. Vettel, *Conversations with a Maverick Scientist.*

14. On Cape, see private papers at ROHO, and Hughes, *Ronald Cape Oral History* (Berkeley, 2004). On Farley, *Alumni Bulletin*, fall 1971, 7, in Cetus Scientific Laboratories, Business Plan, Appendix I, March 1972, issued to Aldo Test, in CACP.

15. Ward to Management Group, 9 Oct. 1972, CACP, box C0120495971, unsorted files. Glaser to Cape, 14 Jan. 1972, CACP, box C0120495971, unsorted files.

16. Hughes, *Alafi Oral History.* Farley to all partners, memo 1, 18 Nov. 1971, CACP, box C0120495971, files unsorted. On Green Cross, CACP, box C0120605569, file: Green Cross; on CRI, Thomas Watterson, "Genetic Engineering Lures Investors With Patience," *The Christian Science Monitor,* 27 Aug. 1981, 11; Peter Daly, *The Biotechnology Business* (Guildford, 1985).

17. Cetus Scientific Laboratories, Inc., Addendum to Business Plan, 5 Apr. 1972, CACP, box C0120495974, file: Business Plans, 101–4.

18. Hughes, *Ron Cape Oral History.*

19. Cetus Scientific Laboratories, Business Plan and Appendix I, March 1972 (copy no. A-13, issues to Aldo Test and reissued to PJF), 2, 10–11, 32–36.

20. Cetus Scientific Laboratories, Business Plan, 4 and 55, in CACP, box C0120495974, file: Certificate of Incorporation.

21. Handwritten note to Ron with a copy of IMS lists attached, no date, CACP, box: C0120495974, file: Business Plan, confirmed in a conversation with Ron Cape, 2 June 2004. Vettel, *Conversations with a Maverick Scientist*; Hughes, *Ron Cape Oral History.* Conversation with David Taft, Menlo Circus Club, 11 Apr. 2004.

22. Memo on 15 Sept. 1972 for 22 Sept. 1972 Board of Dir Mtg, CACP, box C0120495966, file: Board Min, 1972. Farley to Mngmt Group, 20 Sept. 1972, CACP, box C0120495971, file: unsorted.

23. Glaser to everybody else, 7 Dec. 1971, CACP, box C0120495971, file: unsorted; Notice of Meeting to Cetus Board of Directors, CACP, box C0120495966, file: Board Minutes 1973.

24. Glaser to everybody else, 7 Dec. 1971, CACP, box C0120495971, file: unsorted. Glaser to Ron Cape, 14 Jan. 1972, memo 35, CACP, box C0120495971, file: unsorted.

25. Pete to Alafi, Farley, Cape, Glaser, and Ward, 21 Jan. 1972, memo 37, CACP, box C0120495971, file: unsorted.

26. Memo for Board of Dir. mtg., 13 July 1973, CACP, box C0120495966, file: Board Min 1973.

27. Board Meeting, 22 Sept. 1972, CACP, box C0120495966, file: Board Meeting, 1972. Conversation and interview with Ron Cape, 2 June 2004. Ron to MA, PJF, REC, DAG, 3 Jan. 1972, #22, CACP, box C0120495971, file: unsorted.

28. Among many other instances, see Cape to Nowotny, 19 Nov. 1974, CACP, box C0120628365, file: Hoffmann-LaRoche; also, Hughes, *Moshe Alafi Oral History*.

29. Ward to Alafi, Cape, Farley, Glaser, and TM, 21 May 1973, CACP, box C0120495971, files: unsorted. Alafi, Cape memo for Board of Directors meeting, 13 July 1973, CACP, box C0120495966, file: Board Minutes 1973.

30. Cape to Cetus Board, 26 Mar. 1973, CACP, box C0120181581, file: consultants, J. Yule Bogue. Alafi and Cape memo for Board of Dir meeting, 13 July 1973, CACP, box C0120495966, file: Board Min 1973; Vettel, *Conversations with a Maverick Scientist.*

31. Memo for Board of Dir. mtg., 13 July 1973, CACP, box C0120495966, file: Board Min 1973.

32. Glaser to Gordie, 8 Oct. 1973, Tomkins Papers, in UCSFA Archives, MSS 94–14, carton 1, file: Glaser. Farley to Demain, 27 July 1972, CACP, box C0120181581, file: Arnold Demain; Cetus SAB contract with Henry Rapoport, 1 Oct. 1972, CACP, box C0120495970, file RANN.

33. Bogue, Conference on Management Problems in Producing New Medicines, 3 Nov. 1965, CACP, box C0120181581, file: consultants, J. Yule Bogue; A. Demain, "The Industrial Revolution—Microbiologically Speaking," 2003 SIM Annual Meeting Program & Abstracts, p. S132; on Hopwood, see Goulden to Hopwood, 17 Sep. 1974, CACP, box C0120181581, file: David Hopwood, 1970–; on other SABs, see Cape, Glaser, and Alafi oral histories.

34. Lederberg to Ron and Don, 17 Aug. 1974, CACP, box C0120181581, file: consultants: Lederberg; Wright, "Recombinant DNA Technology and Its Social Transformation, 1972–1982," *OSIRIS*, 1986, 2:310.

35. Hughes, *Ron Cape Oral History.*

36. Notes from telephone interview with Jay Groman, 22 Aug. 2001; Cape to All Cetus Empl., 19 Dec. 1973, CACP, box C0120495966, file: Board Minutes, 1973. Mahuron to All Cetus Empl., 28 Nov. 1972, CACP, box C0120495971, no files.

37. The best source of information regarding corporate culture at Cetus comes from the employee newsletter, "Flukes"; benefits initiated by the employees are reviewed in Mahuron to Mngmt Group, 5 Sept. 1972, CACP, box C0120495971, unsorted files; Hughes, *Ron Cape Oral History.*

38. Ron to Don and Cal, 19 Nov. 1971, CACP, box C0120495971, file: unsorted. Telephone conversation with Jay Groman, 22 Aug. 2001.

39. Farley, message to investors, "Nature of Business—General," 3, undated 1973, CACP, attached to Cetus Scientific Laboratories Business Plan, CACP, file: unsorted.

40. This is just a small sample of the problems that plagued Cetus. See Ward to Alafi, Cape, Glaser, Farley, and Mahuron, 15 Feb. 1973, CACP, box C0120495971, file: unsorted; for instance, Berick to Cape, Aiello, Brunner, Chayie, Farley, Geigert, Goulden, Hansen, Mahuron, Miller, Neidleman, and Williams, 17 July 1975, CACP, box C0120409705, file: unsorted.

41. Berick to Brown, Goulden, Raymond, Cape, Farley, and Ward, 1 Nov. 1974, CACP, box C0120409705, file: unsorted. Berick/Hansen to Brown, Goulden, Raymond, Cape, Farley, Glaser, Miller, and Ward, 27 Nov. 1974, CACP, box C0120409705, file: unsorted. Hansen to Ward, 18 Oct. 1972, CACP, box C0120495971, file: unsorted. Goulden to Wagman, Assoc. Dir. of Microbiological Sciences—Antibiotics at Schering Corp., no date, CACP, box C0120565135, file: Schering AG 1973–9; Ron to DG, PF, RC, MA, and CW, 19 Nov. 1971, memo 10, CACP, box C0120495971, file: unsorted; Ron to MA, PJF, REC, DAG, and CBW, 10 Jan. 1972, CACP, box C0120495971, file: unsorted. Ward to Management, 5 Sept. 1972, CACP, box C0120495971, file: unsorted. Berick to Goulden, Miller, Brown, Cape, Farley, Glaser, Hansen, Peterson, Raymond, and Ward, 20 Aug. 1974, CACP, box C0120409705, file: unsorted. Berick to Alafi, Mahuron, Chayie, Glaser, Goulden, Miller, Neidleman, Brown, Cape, Farley, Glaser, Hansen, Peterson, Raymond, Ward, Oldenkamp, and Merrill, 20 Aug. 1974, CACP, box C0120409705, file: unsorted.

42. Cape to Alafi, Farley, Glaser, Mahuron, and Ward, 7 Dec. 1972, CACP, box C0120495971, file: unsorted. *Science*, vol 189, 502 and 21 March, 1052; Stent, "The Ode to Objectivity," *Origins of Molecular Biology* (or, *Atlantic*, November 1971, p. 12: 1979), 231. Cape frequently spoke at, among others, McGill and Princeton University. Among the many receipts, see Farley to Playboy Club Int'l., 11 Feb. 1975, CACP, box C0120495966, file: unsorted.

43. Accid and safety rep, 8 Mar. 1974 and 25 Feb. 1974, CACP, box C0120409705, file: unsorted.

44. Bogue to Ward, 17 July 1974, CACP, box C0120181581, file: consultants, J. Yule Bogue.

45. Bruner to Bio Group, 31 Oct. 1974, CACP, box C0120495971, file: unsorted.

46. Farley to Bogue, 22 July 1974, CACP, box C0120181581, file: consultants, J. Yule Bogue.

47. Alafi to all groups, handwritten memo, 17 Dec. 1973, CACP, box C0120495971, file: unsorted. Safety Report, 25 Feb. 1974, CACP, box C0120409705, file: unsorted. Memo, no author though probably Alafi to Cape, undated, CACP, box C0120409705, file: unsorted. Corresp by Mahuron in CACP, box C0120495971, file: unsorted. Hughes, *Moshe Alafi Oral History*. Vettel, *Conversations with a Maverick Scientist*.

48. Pete (Farley) to MA, PJF, REC, DAG, and CBW, memo 11, 8 Dec. 1971, CACP, box C0120495971, file: unsorted. Memo meeting report, 3, 4, and 5 Jan. 1973, CACP, box C0120605568, file: Stauffer Chemical. Wheaton of Honeywell to Cape, 8 Feb. 1973, CACP, box C0120628365. Farley to MA, PJF, REC, DAG, and CBW, memo 21, 29 Dec. 1971, CACP, box C0120495971, file, unsorted.

49. Ward to Alafi, Cape, Farley, Glaser, and Mahuron, 21 May 1973, CACP, box C0120495971, file: unsorted; interview with Ron Cape, 20 June 2004. Memo on 18 Jan. 1974, re: Board of Dir mtg 25 Jan. 1974, CACP, box C0120505932, file: Board Min, 1974.

50. Ward to Alafi, Cape, Farley, and Glaser, 18 Dec. 1973, CACP, box C0120495971, file: unsorted. Bruner to Ward and Management, 1 Mar. 1973, CACP, box C0120495971, file: unsorted; Goulden to Hopwood, 19 Sept. 1974, CACP, box C0120181581, file: Hopwood, 1970–8.

51. Among many other correspondences, see handwritten note by Cape, "EVERYBODY: Find out about this!" (CACP, box C0120495967, file: cellulose). Lederberg handwritten note to Cape, 17 Aug. 1972, CACP, box C0120181581, file: consultants: Lederberg.

52. Goulden memo re: conversation with Dr. Wilke, 4 Feb. 1975, CACP, box C0120495967, file: cellulose; see also CACP, box C0120495967, memo announcement. Lederberg to Demain, Glaser, Hopwood, Lederberg, Ward, Bio Group, 9 May 1975, re: Hagan, "US is Trying to Do a Lot," *NYT*, 21 Sept. 1975, 8, CACP, box C0120495967, file: Chitin. On cephalosporin, see CACP, box C0120495967, file: Cephalosporin. Schottle and Steuer to Cape, 3 Dec. 1974, and Djerassi to Witzel, 4 Oct. 1974, CACP, box C0120565135, file: Schering AG 1973–9. See any minutes from board meetings in 1974, CACP, box C0120505932, file: Board Minutes, 1974.

53. Swanson telling the partners at Kleiner-Perkins comes from conversation and notes with his roommate and colleague at the time and current partner of KPCB, Brooke Byers, 18 Mar. 2004; Hughes, "Making Dollars Out of DNA; The First Major Patent in Biotechnology and the Commercialization of Molecular Biology, 1974–1980," *Isis*, 2001, 92:541–75; Hughes, *Robert Swanson Oral History*. Hughes, *Moshe Alafi Oral History*.

54. Cape to Glaser, Lederberg, et al., 7 Mar. 1973, CACP, box C0120628365, file: HEW, 1970–9. Bellamy to Cape, 19 Oct. 1973, CACP, box C0120495972, file: Gen Elec, 1973–7.

55. Lederberg to Bob? 19 Nov. 1974, CACP, box C0120181581, file: consultants: Lederberg; Lederberg to Ron Cape, 11 Feb. 1973, CACP, box C0120495972, file: Gen Elec, 1973–7.

56. Advertisement in *Scientific American*, Ron Cape private papers, ROHO, UC Berkeley. Ron and Pete to Glaxo, London, 13 Oct. 1972, CACP, file: Glaxo, 73–9.

57. Hughes, *Ron Cape Oral History*; conversation with Ron Cape, 2 June 2004.

Conclusion

Note to epigraph: Horace Judson, *The Eighth Day of Creation* (New York, 1979), 160.

1. Susan Wright, *Molecular Politics* (Chicago, 1986), 144.

2. Berg to Anderson, 18 Oct. 1974, in SUA, binder: Pre-Asilomar Conference, correspond A–H, 1974–5. Sheldon Krimsky, *Genetic Alchemy* (Cambridge, 1982); Michael Rogers, *Biohazard* (New York, 1977). Berg to Kurt Jacoby, 21 Nov. 1962, in SUA, Paul Berg Papers, SC358, box 1, file "Correspondence, 1960–2; A-J."

3. For a sample of arguments in support of Asilomar, see Paul Berg Papers, SUA, SC358, any correspondence. Master Lord Ashby, FRS to Berg, 4 Nov. 1974; Brown forward to Berg from Dr. Allan Campbell, prof. of biology at Stanford, 14 Nov. 1974; Donald Brown to Type 3 Session Committee, co-signed with Berg, 17 Dec. 1974: all in Paul Berg Papers, SUA, SC358, binder: Pre-Asilomar Conference, correspondence.

4. For a sample of arguments opposing Asilomar, see Paul Berg Papers, SUA, SC358, any correspondence. Irving Crawford to Berg, 18 Nov. 1974, in Paul Berg Papers, SUA, SC358 binder: Pre-Asilomar Conference, correspondence A–H, 1974–5.

5. Bernie David of Harvard Medical School to Berg, 12 Nov. 1975, in Paul Berg Papers, SUA, SC358, binder: Pre-Asilomar Conference, correspondence S–Z, 1974–5. Sally Smith Hughes, "Making Dollars Out of DNA," *Isis*, 2001, 92: 541–74.

6. Because of the inexact nature of practical research, scholars have had to estimate the number of genetic engineering experiments that touch on practical applications. Most scholars believe that the minimum number of genetic

engineering experiments underway in 1975 was 1,440, though in all likelihood, the figure is probably much higher. Susan Wright, "Recombinant DNA Technology," 303–60.

7. Helling to McNiff, 11 Jun. 1992, in Patricia Gossel's files, SI/NMAH—Science in American Life. Mr & Mrs. John Aszkler, Jr., to Berg, 13 Nov. 1974, in Paul Berg Papers, SUA, SC358, binder: Pre-Asilomar Conference, correspondence A–H, 1974–5.

8. "Playing with Genes," *The Economist*, 8 Nov. 1975, 18.

9. Sheldon Krimsky, *Genetic Alchemy* (Cambridge, 1982).

10. Paul Ehrlich to Berg, 6 June 1975, in Berg Papers, SUA, SC358, binder: Pre-Asilomar Conference, correspondence S–Z, 1974–5. Berg to Dyson, 9 June 1975, in Berg Papers, SUA, SC358, binder: Post-Asilomar Conference. Berg to Sydney Brenner, 19 Sept. 1974, in Berg Papers, SUA, SC358, binder: Pre-Asilomar Conference, correspondence A–H, 1974–5.

11. This account of the conference is indebted to Susan Wright, *Molecular Politics* (Chicago, 1986), 145–48. Lederberg to Berg, 10 Mar. 1975 (draft), in the Berg Papers, SUA, SC358, box 13, binder: correspondence 197?, file: Post-Asilomar.

12. Wright, *Molecular Politics*, 153.

13. Wright, *Molecular Politics*, 151. Paul Berg Papers, SUA, SC358, box 1 of 2, newspaper clippings: Ethics of Guidelines for Genetic Research. *Science*, 10 Oct. 19, 135.

14. "Report of the Organizing Committee of the Asilomar Conference on Recombinant DNA Molecules," Berg Papers, SUA, SC358, box 13, binder: Final Report (originals). Hughes, "Paul Berg," 156. Wright, *Molecular Politics*.

15. JMW to Rutter, 5 Feb. 1981, in the Rutter Papers, UCSF Archives, box 2, series II, file: Financial Memos, Materials. Edward Feder to Assist Chancellor Errol Mauchlan, 30 Apr. 1970, in the Papers of the Chancellor for UC, BANC, CU-149, box 115, file: 41, 662 T Biochemistry L&S. Hughes, "Making Dollars Out of DNA," 564.

16. "Industrial Program—Cetus Corporation, February 1986," of Cetus Corporation Annual Report—1986. Oral histories by Hughes on Cohen, Kornberg, Berg, Swanson, and Betlach. Hughes, "Making Dollars Out of DNA," 92. Eyal Press and Jennifer Washburn, "The Kept University," *The Atlantic Monthly*, March 2000, 39–54.

17. Hughes, "Arthur Kornberg: Biochemistry at Stanford, Biotechnology at DNAX," 117–18. Julius Comroe to Dr. W. V. Epstein, 4 June 1970, in Department of the History of Health Sciences, UCSF Archives, AR 87–46, file: 1963, Departmental Projections for Future Development. *Synapse*, 14 Aug. 1975, 1; and "Haight Action, No. 15, August 1971," in UCSF Archives, Neighborhood PR, Haight-Ashbury, Health Community, 1974–5, and Neighborhood PR 1964–1974 (1) 1971. *San Francisco Chronicle*, 2 July 1973; *KPIX* editorial, 27 Jan. 1973; *San Francisco Examiner*, 3 Mar. 1973: all in AR News Services 85–15, Records, 1965–1982. File: Neighborhood PR, Jan.–Feb. 1976.

18. Hughes, "Making Dollars Out of DNA," 565.

19. Cynthia Robbins-Roth, *From Alchemy to IPO* (Cambridge, 2000), xi.

20. Indeed, growth in genetic engineering would continue for some time. For instance, the number of recombinant DNA experiments sponsored by the NIH grew nearly fivefold by the next year while the average equity invested in new genetic engineering firms tripled (Wright, "Recombinant DNA Technology," 320–22). Martin Kenney and Urs von Berg, "Technology, Entrepreneur-

ship and Path Dependency," *Industrial and Corporate Change*, 8:1, 67–103. Bay Area Bioscience Center, "Industrial Records," 2000 Report. Hughes, "Making Dollars Out of DNA," 92; Wright, *Molecular Politics*; and Judith Stein, *Running Steel, Running America* (Chapel Hill, 1998).

21. For instance, in 1968, Nobel prize–winning molecular biologist Jacques Monod stole away from his laboratory at the Pasteur Institute to lead colleagues and students in an all-night battle against the police in the Latin Quarter of Paris; according to Mark Ptashne of the Harvard Biological Laboratories in Cambridge, where more students and faculty joined organizations such as Science for the People than anywhere else in the world, his laboratories' success with repressor-genes was the direct result of his generation's willingness to take "psychic risks"; and seemingly placid University of Tokyo students who had taken the lead in plasmid research—the same field that helped catapult Stanley Cohen's career—participated in larger public demonstrations, singling out in particular the rigid laboratory rituals established by their scientific mentors. Interview transcripts with Tom Kiley, lead counsel for Genentech.

Sources Consulted

Manuscripts and Collections

Pacific Studies Center Archives and Library (Mount View, Calif.)
Rockefeller Archive Center
  The Rockefeller Foundation Papers, RG 1.2, 205D and RF 1.1, 200D
    Holdings:
      Calvin
      Loomis
      Loring
      Stanley
      Van Niel
      Weaver
Smithsonian Institution,
  National Museum of American History, Science and Medicine Division
    Ray Kondratas, Corporate Records File
    Patricia Gossel, Investigator Records File
Stanford University Archives
  Paul Berg Papers, SC358
  Annual Commencement, Order of Exercises
  The Healing Arts, 533 0/4 h
  Arthur Kornberg Papers, SC 359
  Joshua Lederberg Papers, SC 186
  Provost Richard Lyman Papers, SC99
  Medical Center Daily Announcements, 533 c/4
  Medical Center Memo, 533 0/4 m
  Medical Center News Bureau Scrapbooks, 536 0/4 s
  Stanford Daily
  Frederick Terman Papers, SC 0160
  University Bulletin, Directory of Officers and Students
  University Directory
Stanford University Medical School, Lane Library and Archives
  Dean Robert Alway Collections, S1D5
  Directory
University of California, Berkeley, Bancroft Library
  Horace Barker Papers, CU-467
  California Senate and Assembly Bills
  Daily Californian
  Social Protest Collection, 86/157c
  University of California, Chancellor Papers (Strong), CU—149, 695
  University of California, General Catalogue
  University of California, Office of the Registrar

University of California, President's Papers, President's Files, CU-5
University of California, Recorder of the Faculties
University of California, Schedule and Directory
University of California, Statistical Addenda
Wendell Stanley Papers, MSS 78/18 c

University of California, San Francisco, Kalmanovitz Archives and Special Collections
Alumni-Faculty Bulletin, School of Medicine
Announcements of the Medical School (and School of Medicine)
Dean's Office Records, 1936–1989, AR 90–56
Cancer Research Institute
Cardiovascular Research Institute (and AR-46)
Hormone Research Laboratory
Organized Research Units
General Catalogue
Health Sciences, Announcements of the School of Medicine
Administrative Papers, Dept. of the History of Health Sciences, AR 87–46
C. H. Li Papers, MSS 88–9, Series I
William Rutter Papers, MSS 94–54
John Saunders Papers, unsorted files, MSS 90–73
UCSF Magazine
Synapse

University of Chicago, Regenstein Library and Archives, Special Collections
Bulletin of Atomic Scientists
Papers of the Federation of Atomic Scientists

Newspapers and Periodicals: Frequent Citations, 1946–1972

Journal of Biological Chemistry
Journal of Medical Education
Nature
Oakland Tribune
Palo Alto Times (and *Weekly*)
Proceedings of the National Academy of Science
New York Times
San Francisco Chronicle
San Francisco Examiner
San Jose Mercury News

Index